BEYOND THE MOON

SMITHSONIAN HISTORY OF AVIATION
AND SPACEFLIGHT SERIES

Dominick A. Pisano and Allan A. Needell, Series Editors

Since the Wright brothers' first flight, air and space technologies have been
central in creating the modern world. Aviation and spaceflight have trans-
formed our lives—our conceptions of time and distance, our daily routines,
and the conduct of exploration, business, and war. The Smithsonian History
of Aviation and Spaceflight Series publishes substantive works that further
our understanding of these transformations in their social, cultural, political,
and military contexts.

BEYOND THE MOON

A GOLDEN AGE
OF PLANETARY
EXPLORATION
1971–1978

ROBERT S. KRAEMER

SMITHSONIAN INSTITUTION PRESS
WASHINGTON AND LONDON

The color photographs have been reproduced with the generous support of the
Planetary Society, 65 North Catalina Avenue, Pasadena, CA 91106-2301,
www.planetary.org.

Grateful acknowledgment is made for permission to quote "If Only We Had Taller
Been" by Ray Bradbury from *Mars and the Mind of Man* by Ray Bradbury, Arthur C.
Clarke, Bruce Murray, et al. Copyright © 1973 by Harper & Row, Publishers, Inc.
"If Only We Had Taller Been" by Ray Bradbury, copyright © 1973 by Ray Bradbury.
Reprinted by permission of HarperCollins Publishers, Inc.

COPY EDITOR: Karin Kaufman
PRODUCTION EDITOR: Ruth G. Thomson
DESIGNER: Amber Frid-Jimenez

Library of Congress Cataloging-in-Publication Data
Kraemer, Robert S.
   Beyond the moon : a golden age of planetary exploration, 1971–1978 / Robert S. Kraemer.
      p.   cm.
   Includes bibliographical references and index.
   ISBN 1-56098-954-8 (alk. paper)
      1. Space probes—History.   2. Planets—Exploration—History.   3. Space
   sciences—United States—History.   4. Kraemer, Robert S.—Career in space sciences.
   5. United States. National Aeronautics and Space Administration.   I. Title
   TL795.3.K73   2000
   629.43′54′09–dc21   00-0269963

British Library Cataloguing-in-Publication Data available

Manufactured in the United States of America
06 05 04 03 02 01 00   5 4 3 2 1

To the memory of Carl Sagan, dedicated scientist,
outstanding proponent of planetary exploration,
and superb educator of the human race.

CONTENTS

Contents

FOREWORD

NASA AND THE QUEST FOR KNOWLEDGE ABOUT THE PLANETS

Roger D. Launius

Introduction

To some Americans, from the 1950s to the present, space has represented prestige and the American image on the world stage. To others it has signified the quest for national security. To still others space is, or should be, about gaining greater knowledge of the universe. It represents, for them, pure science and the exploration of the unknown. Even so, the history of space science and technology is one of the largely neglected aspects in the history of the space program. This important book by Robert S. Kraemer explores the history of planetary space science efforts throughout the 1970s, an era that has been remembered by those who participated in it as a golden age of planetary science.[1]

Virtually every year of the decade brought the launch of at least one major planetary probe. Several other probes that were not launched until the late 1980s were also started.[2] Indeed, twelve planetary probes launched during the 1970s visited all the planets of the solar system except Pluto, some more than once and for extended exploration, as detailed in the appendix. The solar system exploration program of the

1970s was the stuff of legend and myth in some measure because of its success. Yet it was also much more. It represented a rich harvest of knowledge about Earth's neighboring planets, a transformation of our understanding of the solar system's origin and evolution, and a demonstration of what might be accomplished using limited resources when focusing on scientific goals rather than large human spaceflight programs aimed at buttressing American prestige.[3]

Solar System Exploration before 1970

This large-scale effort of the 1970s did not take place by magic. It required visionary leadership, strong-willed management, and persevering execution. The foundation for this was laid in the 1960s, when space science first became a major field of study. During that decade both the United States and the Soviet Union began an impressive effort to gather information on the planets of the solar system using ground-, air-, and space-based equipment. The studies emanating from these new data revolutionized humanity's understanding of Earth's immediate planetary neighbors. These studies of the planets, perhaps as much even as the Apollo Project, captured the imagination of people from all backgrounds and perspectives. Photographs of the planets and theories about the origins of the solar system appealed to a very broad cross section of the public. As a result, NASA had little difficulty in capturing and holding a widespread interest in this aspect of the space science program.

During the decade of the 1960s, as a direct outgrowth of the Apollo mandate to land Americans on the Moon by the end of the decade, NASA space science focused much of its efforts on lunar missions with Ranger, Surveyor, and the Lunar Orbiter.[4] Even so, a centerpiece of NASA's planetary exploration effort in this era was the Mariner program, originated by NASA in the early part of the decade to investigate the nearby planets. Built by Jet Propulsion Laboratory scientists and technicians, satellites of this program proved enormously productive throughout the 1960s in visiting both Mars and Venus.

Mariner made a huge impact in the early 1960s as part of a race between the United States and the Soviet Union to see who would be the first to reach Venus. That race was not just an opportunity to best the rival in the cold war; scientists in both the United States and the Soviet Union recognized the attraction of Venus for the furtherance of planetary studies. Both the evening and the morning star, Venus has long enchanted humans and all the more so since astronomers realized that it is shrouded in a mysterious cloak of clouds permanently hiding the surface from view. It is also the closest planet to Earth and a near twin to this planet in terms of size, mass, and gravitation.[5]

After ground-based efforts to view the planet in 1961 using radar, which could see through the clouds, and learning among other things that Venus rotated in a retrograde motion opposite from the direction of orbital motion, both the Soviet Union and the United States began a race to the planet with robotic spacecraft. The United States claimed the first success in planetary exploration during the summer of 1962 when Mariner 1 and Mariner 2 were launched toward Venus. Although Mariner 1 was lost during a launch failure, Mariner 2 flew by Venus on 14 December 1962, at a distance of 34,626 kilometers (21,641 miles). It probed the clouds, estimated planetary temperatures, measured the charged particle environment, and looked for a magnetic field similar to Earth's magnetosphere (but found none). After this encounter Mariner 2 sped inside the orbit of Venus and eventually ceased operations on 3 January 1963 when it overheated. In 1967 the United States sent Mariner 5 to Venus to investigate the atmosphere. Both spacecraft demonstrated that Venus was a very inhospitable place for life to exist, determining that the entire planet's surface was a fairly uniform 800 degrees Fahrenheit, thus ending the probability that life—at least as humans understood it—existed on Venus.[6]

At the same time Mars attracted significant attention, an attraction it has yet to relinquish for most planetary scientists, prompting missions there as well. In July 1965 Mariner 4 flew by Mars, taking twenty-one close-up pictures. Mariners 6 and 7, launched in February and March 1969, each passed Mars in August 1969, studying its atmosphere and surface to lay the groundwork for an eventual landing on the planet. Their pictures verified the Moon-like appearance of Mars and gave no hint that Mars had ever been able to support life. Among other discoveries these probes showed that much of Mars was cratered almost like the Moon, that volcanoes had once been active on the planet, that the frost observed seasonally on the poles was made of carbon dioxide, and that huge plates indicated considerable tectonic activity. Mariner 9, scheduled to enter Martian orbit in November 1971, detected a chilling dust storm spreading across Mars; by mid-October dust obscured almost all of Mars. Mariner 9's first pictures showed a featureless disk, marred only by a group of black spots in a region known as Nix Olympia (Snows of Olympus). As the dust storm subsided, the four spots emerged out of the dust cloud to become the remains of giant extinct volcanoes dwarfing anything on Earth. Olympus Mons, the largest of the four, was 480 kilometers (300 miles) across at the base with a crater 72 kilometers (45 miles) wide at the top. Rising 32 kilometers (20 miles) from the surrounding plane, Olympus Mons was three times the height of Mount Everest. Later pictures showed a canyon, Valles Marineris, 4,000 kilometers (2,500 miles) long and 5.6 kilometers (3.5 miles) deep. Later, as the dust settled, meandering "rivers" appeared, indicating that at some time in the past fluid had flowed on Mars. Suddenly Mars fascinated scientists, reporters, and the public.[7]

Proposals for additional Mariner probes were also considered throughout the 1960s but because of budgetary considerations did not fly during the decade. These space probes, as well as others not mentioned here, accumulated volumes of data on the near planets and changed many scientific conceptions that had long held sway.[8]

The Difficult Politics of Planetary Exploration

Although successes in planetary science were real throughout the 1960s and would become even more significant in the 1970s, all was not rosy with the politics of planetary exploration. In many respects the 1960s proved to be a training ground for how to envision, develop, and gain approval for planetary science missions. These political realities were played out throughout the 1970s, and Robert Kraemer discusses some of them in detail in the book that follows. The labyrinth of modern science policy ensures that those engaged in government-funded science must play a savvy game of bureaucratic politics that is at once both insightful and extreme. A variety of strategies arose to succeed at this game. These included keeping individual projects small to avoid serious scrutiny, bringing aboard the project as many scientific disciplines as possible to ensure that everyone had a stake in the effort, developing large partnerships with multifaceted research and educational institutions in numerous congressional districts, and creating international coalitions, to name only a few.

One issue in planetary exploration that was constantly debated, and never fully resolved, was the tradeoffs resulting from the balance of cost, scale, and schedule for space probes. Those engaged in deciding on planetary missions asked whether or not NASA should build a large number and a variety of small, inexpensive probes or consolidate many kinds of experiments onto a few large, expensive spacecraft? Both sides had valid rationales. Small, inexpensive satellites could not accomplish a great deal at any one time and had limited scientific value, but their smallness made them less conspicuous in the political process, and many could be built and flown and thereby overcome the limitations of any one probe. Also, if one or more of the probes failed, the entire planetary program would not suffer as much. Large, costly satellites, on the other hand, were a scientist's (but not an accountant's) dream provided they worked properly, but if any component failed, the returns could be greatly diminished. Such probes also attracted more scrutiny in Washington and had to be astutely managed to ensure funding. Finally, they took much longer to shepherd to completion. It was not uncommon for huge projects to take more than a decade for research, development, and launch.

Between the 1960s and the present, various NASA leaders have swayed back and forth on this question, much of the time advocating, but not always able to deliver, a mixture of large and small spacecraft to avoid the long hiatus that came if a mission failed. Such an approach, although having drawbacks, was designed to minimize the potential difficulties envisioned in a spacecraft's failure. Robert Kraemer discusses this approach as it relates to planning planetary missions during the 1970s.

One overwhelmingly significant issue must be discussed as background for the planetary exploration of the 1970s. In the summer of 1967, even as the technical abilities required to conduct an adventurous space science program were being demonstrated, the planetary science community suffered a devastating defeat in Congress when funding for a Mars lander was canceled. No event was more significant, or cast a longer shadow, in the first quarter century of planetary exploration than the political debacle of losing that mission. It was an enormously important object lesson, and its legacy is everywhere apparent in Kraemer's narrative.

No other NASA effort but the Apollo program was more exciting than the Mars program in the middle part of the 1960s. Mars—so much like Earth and possibly even sustaining life—had long held a special attraction to Americans, and the lander would have allowed for extended robotic exploration of the red planet. A projected $2 billion program, the lander was to use the Saturn 5 launch vehicle being developed for Apollo.

However, scientists lacked consensus on the validity of this Mars exploration initiative. Some were excited and supported the mission; most opposed it as too risky and too expensive. Without that consensus in 1967 and with other national priorities for spending—Great Society social programs, urban unrest, and the war in Vietnam—the Mars lander was an easy target in Congress. It was the first space science project ever killed on Capitol Hill. The NASA administrator, James E. Webb, frustrated by congressional action and infuriated by internal dissension among scientists, stopped all work on new planetary missions until the scientists could agree on a planetary program. As 1968 began, the entire United States planetary exploration program consisted of two Mars flybys scheduled for 1969.[9]

The scientific community learned a hard lesson about the pragmatic, and sometimes brutal, politics associated with the execution of "big science" under the suzerainty of the federal government. Most important, scientists realized that strife within the scientific community had to be kept within the community in order to present a united front against the priorities of other interest groups and other government leaders. Scientists learned that they had to resolve internal differences inside their community, not in complaints to the media or in testimony before Congress. While marshaling support within the scientific community could not guarantee congressional support for a mission, without it virtually any initiative would not be funded. Scientists also learned that although a $750 million program found little opposition at any level, a $2 billion project crossed an ill-defined but real threshold triggering intense competition for those dollars.[10] Having learned these lessons, as well as some more subtle ones, the space science community regrouped and went forward in the latter part of the decade with a trimmed-down Mars lander program, called Viking, that was funded and, as Robert Kraemer notes, eventually provided important scientific data in the mid-1970s.

To avoid future imbroglios, NASA formed a Lunar and Planetary Mission Board and an Astronomy Mission Board to assist in planning future missions and to provide a forum to identify and to resolve differences among the scientists. In 1967 and 1968 space scientists hammered out a mutually acceptable planetary program for the 1970s. Although this program continued to emphasize the exploration of Mars by recommending what became the Viking Project, a scaled-back attempt to make a soft landing on Mars, it also included two Mars orbiters and other initiatives.[11]

In sum, the 1960s saw rapid development both of space science and the technological breakthroughs that made it possible. The result was a radical alteration in the common explanations of the origins and development of the universe. It was a heady environment as funds for space science research rose to about three-quarters of a billion dollars per year in the mid-1960s, more than 17 percent of the NASA budget, and satellite probes and orbiters returned far-reaching data that could be analyzed and incorporated into scientific theories. Although much of the history of the American space program has been predicated on the cold war environment of the latter 1950s and the concomitant competition between the superpowers throughout the 1960s, most but not all of that revolved around human spaceflight initiatives. Space science, perhaps fortunately, was able to use the large Apollo program as political cover to obtain approval for its smaller and exceptionally productive planetary probes. That same strategy worked throughout much of the 1970s; however, as projects became larger, more complex, and more expensive, cover proved harder to obtain. In the 1970s such new planetary science initiatives as Galileo, Cassini, and Magellan each sparked intense debate both inside the science community and in the larger governmental funding process.

Rivalries between Planetary Science and Human Spaceflight

Relations between NASA's Office of Space Science (OSS) and Office of Manned Spaceflight (OMSF) were strained from the beginning of the space age, although an uneasy existence has persisted to the present. Space scientists resented the priorities and media attention enjoyed by the human spaceflight programs, especially Apollo. They complained about the lack of plans or funding in these programs for scientific research in general and about the manner in which planetary science lagged given the budgetary priorities of the piloted spaceflight effort. So intense were the rivalries that when NASA decided to include lunar research in Apollo questions arose as to whether OSS or OMSF should be responsible for it. On 6 September 1966 Robert C. Seamans, NASA's deputy administrator, assigned responsibility for all space science, including that to be performed on piloted spacecraft, to OSS but decreed that the funding be carried in the OMSF budget and then transferred to OSS after congressional approval. This arrangement further exacerbated the tension between OSS and OMSF. The scientific staff of OSS complained that OMSF would not adequately fund scientific work; OMSF engineers complained that OSS scientists cared only about their scientific objectives to the exclusion of the needs of the human spaceflight community. To solve this problem, NASA's associate administrator for space science, Homer E. Newell, created a Manned Space Science Division, staffed it with OSS scientists and OMSF engineers, and required that the head of the division report to him on scientific issues and to the head of OMSF on technical and funding issues. One could argue that these measures led to remarkable scientific returns from Apollo, clearly never envisioned as a science program, but not without a fair measure of controversy and infighting among representatives of these two unique facets of NASA's overall mission.[12]

Similar challenges of negotiating the priorities of space science with the human spaceflight effort occurred during the space shuttle program. On 5 January 1972 President Richard M. Nixon approved the development of the shuttle. Slowly and reluctantly, OSS began to adjust to the use of the shuttle for planetary and other space science activities.[13] OSS modified many of its projects—including its planetary probes to Jupiter, Saturn, and Venus—so that they could be launched aboard the shuttle. Accordingly, after the launch of the Voyagers in 1977, NASA canceled the Titan Centaur launch vehicle program and made plans to phase out the Delta- and Atlas-class expendable launch vehicles. Be cause the performance of the shuttle and its then-planned upper stage were less than that of the Titan Centaur, the cancellation decreased the size of the payload that NASA could send to the planets. Subsequently, to restore this capability, NASA decided to develop a shuttle-compatible version of the Centaur. Later, it also canceled these plans for Centaur, then reinstated them, and finally canceled them for good after the Challenger accident in 1986.[14]

As a result of these decisions the planetary science community spent most of the latter 1970s and 1980s whipsawed between launch vehicle priorities and shuttle perquisites. As only one example of the effect these issues had on the planetary science program, the team responsible for Galileo, NASA's Jupiter probe, spent several frustrating years and many millions of dollars trying to adjust the spacecraft's configuration and trajectory to accommodate the shifting capabilities of the shuttle. Originally scheduled for launch in 1982 as an extended follow-on to the Voyager probe, NASA finally launched Galileo in 1991. Over that period, the cost of Galileo increased by about $1.3 billion. As of May 1988 these increased costs of the space science program amounted to more than $900 million. The Challenger accident delayed this project, and perhaps all the planetary science program, by five to ten years. However, the Challenger accident had one salutary effect: OSS was no longer forced to use the shuttle for all of its launches. It could then purchase expendable launch vehicles from commercial vendors and use the shuttle only when required by the mission.[15]

Scientists, Engineers, and Competing Priorities

One of the fundamental tenets of successful spaceflight was the program management concept, originated in the ballistic missile development program of the 1950s and transferred to NASA almost with the creation of the agency in 1958. This concept posited that successful spaceflight missions depended on maintaining a balance between three critical factors: cost, schedule, and reliability. These three factors were inextricably interrelated and had to be managed as a group. Many also recognized these factors' constancy; if program managers held cost to a specific level, then one of the other two factors, or both of them to a somewhat lesser degree, would be adversely affected. This has held true for virtually every spaceflight program, including planetary science missions. The schedule for these missions was often less firm than human spaceflight activities. So it often took a backseat to cost and reliability. Accordingly, there is an important legacy of slippage in planetary science missions.[16]

Getting all the personnel elements to work together always challenged project managers, regardless of whether or not they were civil service, industry, or university personnel. This was especially true of planetary science missions, in which engineers designing and flying the spacecraft and scientists developing and using scientific instruments competed to ensure that their priorities found expression in the spacecraft. These groups had competing goals and cultures and always enjoyed an uneasy working relationship. As ideal types, engineers usually worked in teams to build hardware that could carry out the functions necessary to fly the spacecraft to its target and to perform its task once it arrived. Their primary goal involved building vehicles that would function reliably within the fiscal resources allocated to the project. Again as ideal types, space scientists engaged in pure research and were more concerned with designing experiments that would expand scientific knowledge about the planetary body to be explored. They also tended to be individualists, unaccustomed to regimentation and unwilling to concede gladly the direction of projects to outside entities. The two groups always contended with each other over a great variety of issues associated with every planetary mission throughout the history of the space age. For instance, the scientists disliked having to configure payloads so that they could meet time, money, or launch vehicle constraints. The engineers, likewise, resented changes to scientific packages added after project definition because these threw their hardware efforts out of kilter. Both scientists and engineers had valid complaints and had to maintain an uneasy cooperation to accomplish all their objectives.

The scientific and engineering communities within NASA, additionally, were not monolithic, and differences among them thrived. Add to these groups the representatives from industry, universities, and research facilities, and competition to further separate scientific and technical areas resulted on all levels. The NASA leadership generally viewed this pluralism as a positive force within the space program, for it ensured that all sides aired their views and emphasized the honing of positions to a fine edge. Competition, most people concluded, made for a more precise and viable space exploration effort. There were winners and losers in this strife; however, and some harbored ill will for years. Moreover, if the conflict became too great and spilled into areas where it was misunderstood, it could be devastating to the conduct of the planetary science program. The head of OSS worked hard to keep these factors balanced and to promote order so that NASA could effectively further knowledge about the solar system, but getting that balance was never an easy task.[17]

As an engineer working within OSS during the 1970s, Robert Kraemer is uniquely qualified to comment on the planetary science program in the decade. He witnessed controversies between scientists and engineers firsthand, worked to ensure that projects gained and kept a place in the NASA budgetary process, and worked closely on the reconfiguration of missions to meet changing priorities, launch schedules, vehicle alterations, and myriad other changes that gave headaches to everyone working in planetary science. Robert Kraemer joined North American Aviation in 1951, after completing his degree in aerospace engineering, and went to work in the division later named Rocketdyne, the preeminent rocket engine development organization in the United States throughout the 1950s. There he helped design engines for the Jupiter, Thor, Atlas, and Saturn rockets.

In the 1960s Kraemer worked for the Aeronutronic Division of Ford Motor Company, serving as chief engineer for space systems and as manager of its Lunar and Planetary Programs. In this capacity he learned firsthand about the challenges inherent in planetary science programs. Charting the delicate course between the requirements of the scientists for their instruments and the rigid confines of technological systems managed by engineers, he worked on some of NASA's most significant missions of the decade. Kraemer moved to NASA Headquarters in 1967 to work on Mars exploration planning efforts, and from 1970 through 1976 he was director of planetary programs. During this period he shepherded to launch many of the illustrious planetary missions that have given rise to the belief that the 1970s was a golden age of planetary science. That task was not without difficulties. Robert Kraemer earned a reputation within NASA for possessing judicious judgment, superior technical skill, and, most important, the ability to convince scientists and engineers of different persuasions to work together, which was never an easy task. He successfully demonstrated that these disparate groups could march at much the same pace, if not the same step, toward much the same goal. Because of this deep background, Kraemer's insider account of planetary science in the 1970s is a welcome addition to the history of space exploration. It compares well with other books relating experiences in space science.[18]

Conclusion

This important book explores many of the central themes affecting space science during the latter half of the twentieth century. It describes and analyzes the conduct of NASA's planetary science program throughout the 1970s, enhancing our understanding of the Mariner, Pioneer, Viking, and Voyager missions. Not so much a part of the "new aerospace history" as a ringing insider account of the subject, Kraemer's discussion provides the details that only a participant would know about the progress of these important missions.[19] It is a heroic story in Kraemer's telling, one filled with men and women of good character striving to achieve important objectives. They did not always agree with one another and competed ferociously for their respective positions, but they all respected one another. The result was astonishing, and that alone may have given rise to the belief that the planetary missions executed during the 1970s were a golden age of the program. Although the case may be well made for this assessment at present, and Kraemer makes it well, what will be the conclusion of those one hundred years hence? Will probes sent to the planets serve, as Columbus did with the Americas, as vanguards of sustained exploration and settlement? Or will they prove to be more like Leif Ericksson's voyages, stillborn in the public conception of new lands? No one knows at present, but books such as this make a subtle case for many additional voyages of discovery.

Notes

1. There are some fine histories of this subject; however, most are of a popular nature. Some of the better works include William E. Burrows, *Exploring Space: Voyages in the Solar System and Beyond* (New York: Random House, 1990); Henry S. F. Cooper, *The Evening Star: Venus Observed* (New York: Farrar, Straus, and Giroux, 1993); Henry S. F. Cooper, *Imaging Saturn: The Voyager Flights to Saturn* (New York: Holt, Rinehart, and Winston, 1981); Henry S. F. Cooper, *The Search for Life on Mars: Evolution of an Idea* (New York: Holt, Rinehart, and Winston, 1980); Steven J. Dick, *The Biological Universe: The Twentieth Century Extraterrestrial Life Debate and the Limits of Science* (New York: Cambridge University Press, 1996); Edward Clinton Ezell and Linda Neuman Ezell, *On Mars: Exploration of the Red Planet, 1958–1978*, NASA SP-4212 (Washington, D.C.: GPO, 1984); Bevan M. French and Stephen P. Maran, eds., *A Meeting with the Universe: Science Discoveries from the Space Program*, NASA Educational Publication-177 (Washington, D.C.: GPO, 1981); Paul A. Hanle and Von Del Chamberlain, eds., *Space Science Comes of Age: Perspectives in the History of the Space Sciences* (Washington, D.C.: Smithsonian Institution Press, 1981); Jeffrey Kluger, *Journey Beyond Selene: Remarkable Expeditions Past Our Moon and to the Ends of the Solar System* (New York: Simon and Schuster, 1999); Clayton R. Koppes, *JPL and the American Space Program: A History of the Jet Propulsion Laboratory* (New Haven, Conn.: Yale University Press, 1982); Bruce C. Murray, *Journey into Space: The*

First Three Decades of Space Exploration (New York: W. W. Norton,
1989); Homer E. Newell, *Beyond the Atmosphere: Early Years of Space
Science,* NASA SP-4211 (Washington, D.C.: GPO, 1980); and Robert
Reeves, *The Superpower Space Race: An Explosive Rivalry through the
Solar System* (New York: Plenum Press, 1994).

2. There were launches in 1971, 1972, 1973, 1974, 1975, 1976, 1977, and
   1978.
3. Indeed, Representative George E. Brown Jr. (D-CA) remarked in a speech
   to the National Academy of Sciences in 1992 that "[i]t is also important
   to recall that some of our proudest achievements in the space program
   have been accomplished within a stagnant, no growth budget. The development of . . . the Viking lander, Voyagers I and II, Pioneer Venus, . . .
   were all carried out during the 1970s when the NASA budget was flat. It
   would be wise to review how we set priorities and managed programs
   during this productive time" (George E. Brown, Remarks, 9 February
   1992, copy in NASA Historical Reference Collection, NASA History
   Division, Washington, D.C.).
4. On lunar science see these major histories: R. Cargill Hall, *Lunar Impact: A History of Project Ranger,* NASA SP-4211 (Washington, D.C.:
   GPO, 1977); Bruce K. Byers, *Destination Moon: A History of the Lunar
   Orbiter Program,* NASA TM X-3487 (Washington, D.C.: GPO, 1977);
   Linda Neuman Ezell, comp., *NASA Historical Data Book, Vol. II: Pro-
   grams and Projects, 1958–1968,* NASA SP-4012 (Washington, D.C.:
   GPO, 1988), 325–31; David M. Harland, *Exploring the Moon: The
   Apollo Expeditions* (Chicester, England: Wiley-Praxis, 1999); and W.
   David Compton, *Where No Man Has Gone Before: A History of Apollo
   Lunar Exploration Missions,* NASA SP-4214 (Washington, D.C.: GPO,
   1989).
5. Jet Propulsion Laboratory, *Mariner-Venus 1962: Final Project Report,*
   NASA SP-59 (Washington, D.C.: GPO, 1965), 4–6.
6. The story of Venusian radar imagery can be found in Andrew J. Butrica,
   To See the Unseen: A History of Planetary Radar Astronomy, NASA
   SP-4218 (Washington, D.C.: GPO, 1996), 27–53. The story of the
   Mariner mission is relayed in JPL, *Mariner-Venus 1962;* JPL, *Mariner,*
   102–12; and Reeves, *Superpower Space Race,* 177–83.
7. See William K. Hartman and Odell Raper, *The New Mars: The Discoveries of Mariner 9,* NASA SP-337 (Washington, D.C.: GPO, 1974).
8. Robert B. Leighton, Bruce C. Murray, Robert P. Sharp, J. Denton Allen,
   and Richard K. Sloan, "Mariner IV Photography of Mars: Initial Results," *Science* 149 (6 August 1965): 62730.
9. For a discussion of this political debate see Ezell and Ezell, *On Mars,*
   83–120.

10. The average percentage of the total NASA appropriation allotted for space science in 1959–68 was 17.6 percent; it was 17 percent in 1969–78. By 1996 it was less than 10 percent.

11. NASA, Office of Space Science and Applications, "Planetary Program Review," 11 July 1969, copy in NASA Historical Reference Collection, NASA History Division, Washington, D.C.; John E. Naugle, "Goals in Space Science and Applications," *Nuclear News,* January 1969.

12. For example, see A. A. Levinson, ed., *Proceedings of the Apollo 11 Lunar Science Conference, Houston, Texas, January 58, 1970,* vol. 1 (New York: Pergamon Press, 1970), 154; Don E. Wilhelms, *To a Rocky Moon: A Geologist's History of Lunar Exploration* (Tucson: University of Arizona Press, 1993); Paul D. Spudis, *The Once and Future Moon* (Washington, D.C.: Smithsonian Institution Press, 1996); and Curator for Planetary Materials, Johnson Space Center, "Top Ten Scientific Discoveries Made during Apollo Exploration of the Moon," 28 October 1996, NASA Historical Reference Collection, NASA History Division, Washington, D.C.

13. Homer E. Newell, Associate Administrator, NASA, to Dr. [James] Fletcher, Administrator, NASA, "Relations with the Science Community and the Space Science Board," 3 December 1971, NASA Historical Reference Collection, NASA History Division, Washington, D.C.

14. For a discussion, with documents concerning the expendable launch vehicle situation, see John M. Logsdon, gen. ed., with Ray A. Williamson, Roger D. Launius, Russell J. Acker, Stephen J. Garber, and Jonathan L. Friedman, *Exploring the Unknown: Selected Documents in the History of the U.S. Civil Space Program, Vol. IV, Accessing Space,* NASA SP-4407 (Washington, D.C.: GPO , 1999).

15. U.S. General Accounting Office, "Space Exploration: Cost Schedule and Performance of NASA's Galileo Mission to Jupiter," 27 May 1988, GAO/NSIAD88138FS; U.S. General Accounting Office, "Space Exploration: Cost Schedule and Performance of NASA's Ulysses Mission to the Sun," 27 May 1988, GAO/NSIAD88129FS; U.S. General Accounting Office, "Space Exploration: Cost, Schedule and Performance of NASA's Magellan Mission to Venus," 27 May 1988, GAO/NSIAD88130FS; U.S. General Accounting Office, "Space Science: Status of the Hubble Space Telescope Program," 2 May 1992, GAO/NSIAD88118BR; copies available in NASA Historical Reference Collection, NASA History Division, Washington, D.C.

16. Aaron Cohen, "Project Management: JSC's Heritage and Challenge," *Issues in NASA Program and Project Management,* NASA SP-6101 (Washington, D.C.: GPO, 1989), 7–16.

17. Howard E. McCurdy, *Inside NASA: High Technology and Organiza-*

tional Change in the U.S. Space Program (Baltimore, Md.: Johns Hopkins University Press, 1993), 11–98.

18. Newell, *Beyond the Atmosphere;* Wilhelms, *To a Rocky Moon;* Elbert A. King, *Moon Trip: A Personal Account of the Apollo Program and Its Science* (Houston, Tex.: University of Houston, 1989); James A. Van Allen, *Origins of Magnetospheric Physics* (Washington, D.C.: Smithsonian Institution Press, 1983); Spudis, *Once and Future Moon;* and Murray, *Journey into Space.*

19. I have discussed the new aerospace history and its characteristics in other venues. See Roger D. Launius, "National Aerospace Policy and the Development of Flight in America," in *1998 National Aerospace Conference: The Meaning of Flight in the 20th Century, Conference Proceedings,* ed. Tom D. Crouch and Janet R. Daly Benarek (Dayton, Ohio: Wright State University, 1999), 280–85; and Roger D. Launius, ed., *Innovation and the Development of Flight* (College Station: Texas A&M University Press, 1999), 14–15.

ACKNOWLEDGMENTS

I join the human race in thanking the dedicated men and women who accomplished the extraordinary and historic exploration of the solar system during what is now recognized as the golden era of planetary exploration. I have only been able to include a few of my personal "superheroes" in this book; my apologies to the many others equally deserving of recognition for their efforts.

As an engineer myself, I have focused primarily on space engineers and scientists, to the neglect of other essential contributors to the golden era. In the Planetary Programs office at NASA Headquarters, for example, the place only functioned because of efficient, cooperative, and innovative secretaries and administrative staff. As director of the office, I could always rely on Peter Hatt for administrative support, and I was blessed with exceptional directorate secretaries in Mary Ann Gaskins and Danalee Green, both of whom went on to NASA senior management positions. I counted on them not only to lead and efficiently assign our staff of secretaries but also, more important, to properly divert and redirect 97 percent of the overwhelming government paperwork requests so that only the essential 3 percent came to my desk. Without them and our other exceptional secretaries, such as the dynamic Diane Mangel, we engineers and planetary scientists never would have been able to focus on our real job of selling new planetary exploration missions.

I was encouraged to write this book by several professionals, including Prof. James Langford, head of the University of Notre Dame Press; author Terry Cline; professor-author Sister Mary Gerald Leahy; and Dr. Roger Launius, NASA chief historian. Roger in turn put me in contact with Von Hardesty and Mike Gorn, senior editors of the Smithsonian History of Aviation and Spaceflight Series, and Mark Hirsch, acquisitions editor at the Smithsonian Institution Press. It was the strong encouragement and constructive advice from these experienced experts that kept me at my keyboard. And speaking of keyboard, my thanks to my sons Stephen and David for keeping my old word processor functioning, and special thanks to David for help on the graphics.

Assistance in locating illustrations for the book came from space veteran Jurrie van der Woude at the Jet Propulsion Laboratory in Pasadena, California, and from Lynn Albaugh at the Ames Research Center, Moffett Field, California. David DeVorkin and Michael Neufeld of the Smithsonian's National Air and Space Museum provided a very helpful historical critique of chapter 1. Karin Kaufman did a skilled and meticulous job of copy editing, and production expertise came from Mark Gatlin, Ruth Thomson, and the staff at the Smithsonian Institution Press.

Finally, I wish to thank my lovely and spirited Irish-descended wife, Anne, for taking such loving care of my well-being and especially for letting me concentrate at my word processor for long periods without the conversation she craves so dearly. Love you, Anne.

ABBREVIATIONS

ABL	Automated biological laboratory	BOB	Bureau of the Budget
		BOMI	Bomber Missile
ABMA	Army Ballistic Missile Agency	Caltech	California Institute of Technology
AEC	Atomic Energy Commission	CCS	Computer Command Subsystem
AIAA	American Institute of Aeronautics and Astronautics	Code S	Office of Space Science and Applications
AO	Announcement of opportunity	Code SL	Lunar and Planetary Programs Division
APA	Allowance for program adjustment	Code SP	Space Physics and Astronomy Programs Division
APMT	Advanced Planetary Missions Technology	CPAF	Cost plus award fee
		DoD	Department of Defense
ARPA	Advanced Research Projects Agency	DSN	Deep Space Network
		ESA	European Space Agency
ARS	American Rocket Society	GC-MS	Gas chromatograph–mass spectrometer
AU	Astronomical unit		
BMFT	Bundesminister für Forschung und Technologie	GISS	Goddard Institute for Space Science

GSOC	German Space Operations Center	NAS	National Academy of Sciences
GSFC	Goddard Space Flight Center	NASA	National Aeronautics and Space Administration
HUD	Housing and Urban Development	NRL	Naval Research Laboratory
ICBM	Intercontinental ballistic missile	OAST	Office of Aeronautics and Space Technology
IITRI	Illinois Institute of Technology's Research Institute	OMB	Office of Management and Budget
IR	Infrared	OMSF	Office of Manned Spaceflight
IRBM	Intermediate-range ballistic missile	OPWG	Outer Planets Working Group
IRIS	Infrared interferometer spectrometer	OSSA	Office of Space Science and Applications
IRTF	Infrared Telescope Facility	OST	Office of Science and Technology
ISS	International Space Station	P&W	Pratt and Whitney
IUS	Inertial upper stage	PI	principal investigator
JATO	Jet-assisted takeoff	PSAC	President's Scientific Advisory Committee
JOP	Jupiter Orbiter-Probe		
JPL	Jet Propulsion Laboratory	PSC	Physical Sciences Committee
J-S-P	Jupiter to Saturn to Pluto	R&D	Research and development
J-U-N	Jupiter to Uranus to Neptune	RFNA	Red fuming nitric acid
		RFP	Request for proposal
KSC	Kennedy Space Center	ROBO	Rocket Bomber
LeRC	Lewis Research Center	RTG	Radioisotope thermoelectric generator
LPMB	Lunar and Planetary Missions Board	SAR	Synthetic aperture radar
LST	Large Space Telescope	SEP	Solar electric propulsion
MJS	Mariner Jupiter Saturn	SFOF	Space Flight Operations Facility
MJU	Mariner Jupiter Uranus		
MSFC	Marshall Space Flight Center	SSB	Space Science Board (later Space Studies Board)
MVM	Mariner Venus Mercury		
NACA	National Advisory Committee for Aeronautics	SSG	Science Steering Group
		SSSC	Space Science Steering Committee

Abbreviations

SST	Supersonic transport	ULO	Unmanned Launch
STAR	Self-test and repair		Operations
SURMEC	Surface measuring	USAF	U.S. Air Force
	capsule	UVS	Ultraviolet spectroscopy
TOPS	Thermoelectric Outer	VfR	Verein für
	Planets Spacecraft		Raumschiffahrt
TRW	Thompson Ramo		
	Woolridge		

INTRODUCTION

It was one of those rare moments in time. In the brief span of eight years, from 1971 to 1978, Americans launched a fleet of robot spacecraft on paths to the far corners of the solar system, examining close up every planet except remote and tiny Pluto, undertaking what must surely be the greatest burst of exploration in the history of mankind. Immediately on the heels of the Apollo landing on the Moon, every single year from 1971 to 1978 one or more exquisitely engineered spacecraft were launched into deep space from Cape Canaveral, Florida. There were twelve spacecraft in all, and every mission was a stunning success, full of surprising revelations about our neighboring planets and their rings and satellites. Consider briefly just a few of the results.

Launched in 1971, Mariner 9 was the first spacecraft to orbit another planet. It mapped the surface of Mars, revealing volcanoes up to 21.7 kilometers (70,000 feet) high, a Grand Canyon nearly 4,800 kilometers (3,000 miles) long, and clear evidence of massive flows of water. The mission revived and heightened speculation that life could have evolved on Mars.

Pioneer 10, launched in 1972, proved that a spacecraft could survive a passage through the Asteroid Belt beyond Mars. It survived intense radiation to make the first close flyby of gigantic Jupiter and then went on to become the first manmade object to escape the solar system.

Pioneer 11, which was launched in 1973, flew by Jupiter and then on to make the first visit to Saturn, paving the way for the later Voyager 2 Grand Tour mission before heading out of the solar system.

Launched in 1973, Mariner 10 produced the first closeup images of Venus, revealing classic Hadley circulation in the clouds, and then, in the first use of gravity assist, made the first visit to Mercury (followed by two more revisits). The craft's high-resolution mapping revealed Mercury's surface to be heavily cratered like the Moon.

As part of a joint U.S.-German project, Helios 1, launched in 1974, made the closest ever approach to the Sun.

Launched in 1975, Vikings 1 and 2, the most advanced robot machines ever built, mapped Mars from orbit and accomplished the first two successful landings on another planet. The Viking Landers functioned beautifully but found no signs of life, intense ultraviolet radiation apparently preventing the evolution of complex molecules, at least on the surface.

Helios 2, launched in 1976, traveled more than 70 percent of the distance toward the Sun, returning detailed measurements of solar properties over the cycle from solar minimum to solar maximum.

Launched in 1977, Voyager 1 returned high-resolution views of Jupiter and its Galilean satellites, discovering many active sulfur-spewing volcanoes on satellite Io. It also returned closeups of Saturn and its rings and satellites, including Titan, with its thick nitrogen and methane atmosphere.

Voyager 2, launched in 1977, discovered rings around Jupiter and then utilized gravity assist at Saturn to continue with its Grand Tour to the first flybys of Uranus and Neptune. It returned astounding images of satellites, including tortured terrain on Miranda and ice-spewing geysers on Triton. The mission has been hailed as the greatest single voyage of exploration of all time.

Finally, Pioneer Venus 1 and 2, launched in 1978, produced major advances in the understanding of Venus through long-term study from orbit and multiple probes into its thick, hot atmosphere laced with sulfuric acid.

Twelve straight missions of discovery in quick succession and every one fully successful and returning spectacular and surprising revelations. Surely those eight years deserve their common designation as a golden age or golden era of planetary exploration. Just the first phase of a continuing, intense exploration of new worlds that could tell us so much about how Earth and the solar system have evolved, right? Not so. It is an amazing fact that no more planetary missions were launched by Americans for the next eleven years.

How did this historic burst of pioneering activity come to be? It may seem to the casual observer to have been smooth and relatively easy— the National Aeronautics and Space Administration (NASA) proposed missions, the White House requested funding from Congress, Congress approved, and all the spacecraft were built and launched on schedule and generally within their budgets. Nothing in the space program has ever come that easily, however. In this book I will give you a look at some of the struggles behind the scenes: the sometimes frantic efforts of space scientists to sell their particular fields of research; their "dirty tricks" and attempts to shoot down rival mission candidates; the fierce competition to gain priority within the National Academy of Sciences (NAS) and within NASA Headquarters; the frequently frustrating attempts to generate interest and enthusiasm among the "bean counters" of the White House Bureau of the Budget (BOB) and its successor, the Office of Management and Budget (OMB); the "educating" of congressional staffers (no lobbying by NASA allowed, remember); the struggles with technical, management, and budget problems during development of advanced hardware and software; and the sweating out of frequently Perils-of-Pauline troubles with the launch vehicles and spacecraft. It probably won't come as any great surprise that the behavior of the people involved was often even more complex and challenging than the technical problems that had to be overcome.

The challenge of trying to do something difficult, something no one has done before, has always attracted a special class of engineer and scientist. To these bold individuals the thrill of a major "first" far outweighs any fear of failure. As unexpected problems threaten the very life of a spacecraft in deep space you can count on these people to work feverishly around the clock to save the machine of their creation on its pioneering mission. There is no way you could pay people to work with that intensity. It is the thirst to be the first to view new worlds and perhaps find other forms of life that drives them to such effort. As we look behind the scenes of the golden era, I will highlight a few of these bold pioneers. In fact, if it were not for overuse of the familiar words from the *Star Trek* television series, I would have titled this book *To Boldly Go,* for that is indeed what they did. It was my great good fortune to get to know and to work with these pioneers—my personal heroes of the early days of rocketry and the subsequent development of sophisticated robot spacecraft.

Because much of the behind-the-scenes story, especially on the selling of missions, is only recorded in the memories and personal notes of individuals such as myself who were directly involved, I am afraid you will have to tolerate my frequent use of the first person in the narrative, and you are going to get my strictly personal assessment of people. Unfortunately, there is not sufficient room in this book to discuss anywhere near all of the key engineers and scientists. Repeated success was only achieved through the concerted efforts of a cohesive team made up of many dedicated individuals (contractor, university, and NASA personnel) who devoted years of their lives to accomplish these fantastic missions of exploration.

I will discuss mission results only briefly, as the project scientists on all of the planetary missions either coauthored or edited excellent summaries of the scientific results and they are far superior to anything I might attempt. An up-to-date overall summary of mission results can be found in the *Encyclopedia of the Solar System,* published by Academic Press in 1999, and I have listed other appropriate publications in the references.

Will we ever see another golden era of planetary exploration? Are there lessons to be learned from the missions of the 1970s? What can planetary scientists and NASA managers do to create another great burst of knowledge about our solar system? In the epilogue I will analyze past and recent history and then gaze into my crystal ball.

ON THE SHOULDERS OF GIANTS

DEVELOPING THE TECHNOLOGY

The golden era of planetary exploration did not start out so golden. America's efforts to explore the solar system began with eleven straight mission failures. The Soviet Union, off to such a brilliant start in Earth orbit, had many failures (nineteen by my count) before it launched a truly successful mission to another planet. There was much new technology to be developed, experience to be earned the hard way, more than a few scapegoats to be fired, and long hours of dedicated effort to be devoted to this new frontier of exploration before the rewards were to start pouring in.

There is no way to condense into one chapter the entire history of rocketry and spaceflight, but I would like to highlight and pay tribute to a few of the key individuals who pioneered the technological advances in rocket propulsion, launch vehicles, and spacecraft that made possible the great exploration payoff of the 1970s. They are some of the giants on whose shoulders the planetary explorers stand.

The first challenge of technology was to go beyond the atmosphere-limited capabilities of balloons and aircraft. Various proposals to reach outer space, such as the giant cannon proposed by Jules Verne, were

5

ruled out by the excessive heating and drag that would be encountered in passing through Earth's thick atmosphere. To go into space, completely leaving our atmosphere, one must use rocket propulsion. Although that is commonly accepted wisdom today, it was not always so. An infamous editorial in the *New York Times* on 13 January 1920 dismissed a pioneering technical paper[1] on the feasibility of rocket-powered spaceflight by Dr. Robert H. Goddard, then a physics professor at Clark University in Worcester, Massachusetts, stating sarcastically that Goddard "seems to lack the knowledge ladled out daily in high schools"; that is, that anyone with even a high school knowledge of physics knew that a rocket needed air to push against to develop any thrust. The *Times* was wrong (although the paper didn't formally retract its statement until Apollo 11 was ready to land on the Moon in 1969). A rocket is a reaction device. By accelerating exhaust gas it creates a reaction force in the opposite direction, and because gases can be accelerated to even higher velocities in a vacuum than in the atmosphere, a rocket engine is actually more efficient in space than at sea level.

The first person credited with putting the rocket reaction process into solid technical terms was Konstantin Tsiolkovsky, a mathematics teacher from the remote Russian village of Kaluga who developed the basic mathematics of rocket-propelled spaceflight in the late 1800s and documented it in a milestone paper in 1903.[2] His theoretical work was further advanced in the 1920s by America's Robert Goddard and by Romanian-born Hermann Oberth.[3]

Liquid-Rocket Pioneers

Solid-propellant rockets using black powder were invented by the Chinese way back in the twelfth century. Solid propellants, though, are inherently limited in chemical energy. It was going to take higher-energy liquid propellants to venture into deep space. Starting in the early 1920s Goddard went beyond theory to experimental work in the development of liquid rockets, which obtain their energy from the combustion of a liquid oxidizer and a liquid fuel. For his oxidizer Goddard elected to use liquid oxygen, which had the inconvenience of being cryogenic but was nontoxic and burned well with a variety of common fuels, such as ammonia, alcohol, and gasoline.

On 16 March 1926 a Goddard rocket flew for two and a half seconds. Burning liquid oxygen and gasoline, it crabbed sideways for 56 meters (184 feet) and reached an altitude of only 12.5 meters (41 feet). Not overly impressive, but it was still the world's first flight of a liquid-propellant rocket. If you visit the immensely popular National Air and Space Museum in Washington, D.C., you will see a faithful reproduction of this first successful liquid rocket, a spindly thing, 3 meters (9.8 feet) tall, looking like little more than a Fourth of July skyrocket. It featured a small thrust chamber (the combination of a combustion chamber, where the propellants are burned, and a nozzle to expand the resulting hot gases to supersonic velocity) about .6 meters long mounted in the nose, with small, pointed propellant tanks mounted behind, where they were immersed in the hot rocket exhaust gases. With a net thrust of only 40 newtons (9 pounds), it managed to fly, but Goddard quickly realized that this "pull" rather than "push" arrangement complicated rather than helped steering problems, and all of his later rockets put the thrust chamber in the tail.

Goddard made steady, if not rapid, technical progress. In 1935 his rockets attained speeds of more than 1,130 kilometers (700 miles) an hour and an altitude of 2.3 kilometers (7,500 feet). Standing next to Goddard's first liquid-propellant rocket in the Air and Space Museum is one of his last rockets. At 6.7 meters (22 feet) tall and with a thrust of 4,400 newtons (985 pounds), this 1941 rocket is most impressive and incorporates such advanced features as a streamlined shell, turbopumps to pressurize the propellants, aerodynamic fins, and gyroscopes to guide the vehicle via jet vanes moving in the rocket exhaust.[4]

I regret that I never got to meet Dr. Goddard, who died in 1945, although I had numerous conversations in the 1960s and 1970s with his wife, Esther, who continued her interest in spaceflight and was a regular attendee at the annual Goddard Memorial Dinner sponsored by the National Space Club and at functions at NASA's Goddard Space Flight Center (GSFC). In the early 1940s as a high school student I was not aware of Goddard's work, but I had seen movies of small rocket-motor firings by a group who called themselves the American Rocket Society

(ARS). To me that was exciting stuff. I had already concluded that space would be the next great frontier and that I wanted to be an engineer in that new and challenging field. In 1946 the class ahead of me in the Aeronautical Engineering Department at the University of Notre Dame built a small rocket motor using ammonia and gaseous oxygen as propellants. I invited my roommate, Chuck Mason, to join me in observing the first test firing. The countdown 5, 4, 3, 2, 1 was followed by a great hiss and blast of gas out of the small rocket. Chuck remarked at how loud it was, and I knowingly told him yes, that was how rockets sounded. It was more than a little embarrassing to then find out that the propellants had not ignited—it was only the gas discharge we had heard. So much for Bob Kraemer the rocket expert.

Our favorite aero-engineering professor at Notre Dame was Dr. Vincent Goddard. I did not know it at the time, but Vince was a nephew of Robert Goddard. Two of our small class of fourteen aero-engineers at Notre Dame went on to pursue careers in rocketry—myself and close friend Tim Hanrahan, who after graduation managed Corporal missile firing tests at White Sands, New Mexico, and then was for many years the Aerospace Corporation manager for air force launches at Cape Canaveral. As for myself, in 1950 I went from Notre Dame to the graduate school at the California Institute of Technology (Caltech), taking every course they had that was in any way related to rocket propulsion and getting to know people at the nearby Caltech-managed army lab known as the Jet Propulsion Laboratory (JPL), where they were performing firing tests with small liquid-propellant rocket motors. Thus I began a long and rewarding relationship with JPLers, my engineering career seeming to always intertwine with endeavors at JPL.

In the final years of World War II a great advance in rocket technology came out of Germany. Back in the 1920s and 1930s a group of spaceflight enthusiasts in Germany known as the Verein für Raumschiffahrt (Society for Space Travel), VfR for short, had started experimental work with liquid rockets. An enthusiastic teenage member of the VfR was Wernher von Braun, who joined about 1928. He, along with Hermann Oberth, Willy Ley, and other members of the VfR, were confident that given some financial support they could build much larger liquid-fueled rockets. In 1932 they found a receptive audience in

German army colonel Karl Becker and his assistant, Capt. Dr. Walter Dornberger, who was in charge of powder rockets for the German army. Dornberger made the bold move of hiring the young von Braun as his civilian technical director to develop liquid rockets.

The story of their successful development of the V-2 ballistic missile and its use to bombard London in the closing months of World War II has been well covered by historians (and at least one motion picture), so I will not repeat it here. What is pertinent to this book is that the V-2, which was a great leap forward in rocket technology, had many of the features of Goddard's most advanced rockets but scaled up to a much higher level of performance. Whereas Goddard's largest rockets developed a thrust of less than 4,500 newtons, the V-2s generated 250,000 newtons (56,000 pounds) of thrust at liftoff.

Launch Vehicle Progress

As the war in Europe ended, von Braun and his team of key engineers and scientists agreed that they wanted to be captured by the Americans rather than the Russians, or even the British. With a little bit of luck and the quick action of a very alert U.S. Army group under Col. Holger Toftoy, they succeeded in this, and 115 of von Braun's hand-picked crew and thirteen thousand kilograms (fourteen tons) of their documents were whisked off to the United States along with sufficient parts to assemble dozens of V 2s. The army moved them to Fort Bliss in Texas and asked von Braun to oversee the assembly and flight testing of the captured V-2s at the White Sands Proving Ground. In 1950 the army moved the expatriated Germans, along with a number of American engineers, to the Redstone Arsenal in Huntsville, Alabama. In 1956 the group was named the Army Ballistic Missile Agency (ABMA) and then in 1960 was folded into NASA as the Marshall Space Flight Center (MSFC).

After receiving my master of science degree from Caltech in 1951 I joined the rocket propulsion group (later known as Rocketdyne) at North American Aviation, and it was there that I first began to work with von Braun and his team of German engineers. One of the German's senior members, Walter Riedel, I believe known as "Papa" to the

V-2 team, was assigned to assist us at North American. He was a stereotype of the Prussian personality, arrogant and sure that all rocket engines must be designed just like the V-2 engine, which proved to be much more complex than it needed to be (more on that later). He even tried to tell us our English pronunciation was all wrong. Riedel was no help to us at all. We came to realize why von Braun did not especially want him in Huntsville.

I also worked with General Dornberger (the V-2 work got him promoted from colonel to general), who had been assigned to work with Bell Aircraft in Buffalo, New York, and was leading Bell's ambitious manned military orbiter project called variously BOMI (Bomber Missile), ROBO (Rocket Bomber), and DynaSoar. We at Rocketdyne were providing rocket-engine designs for these proposed orbiting vehicles. Dornberger was of only average height but solidly built, and he radiated energy and strong will. It was easy to see how he had been an important force behind the development of the V-2.

The outstanding person on the German V-2 team was the team's technical leader, Wernher von Braun. Tall, handsome, square-jawed, and fair-haired, he was an impressive figure. I fully agree with others who have described him as "charismatic" and as possessing "prodigious quantities of charm, tact, intellect, and leadership ability."[5] Rocketdyne became the favorite engine contractor for von Braun and his ABMA crew because it was following the direction set by the V-2. While Rocketdyne's chief competitor, Aerojet Corporation, pursued the use of storable (noncryogenic) oxidizers such as red fuming nitric acid (RFNA), which proved to be very difficult to work with, Rocketdyne elected to follow up on the V-2's use of liquid oxygen. It proved a wise choice, as oxygen was clean to handle and combusted well with a variety of fuels, including alcohol, gasoline, jet fuel, and hydrogen.

Rocketdyne's first new large rocket engine was an Americanized (simplified for easier production) version of the V-2 engine uprated to seventy-five thousand pounds of thrust. (American engines were identified by pounds of thrust rather than newtons.) Its first application was to boost the U.S. Air Force (USAF) Navaho 1 cruise missile, but it is not surprising that von Braun and the ABMA adopted it to propel their new army Redstone ballistic missile. It was one of these Redstones that was selected to send Alan Shepard on America's first manned suborbital flight. I remember when that particular Redstone engine was being assembled at Rocketdyne's plant in Canoga Park, California. The shop had put a tag on it saying that a human was going to ride with that engine. The Redstone thrust chambers had lots of welds and were rather dingy and unsightly, but the one for Alan Shepard came out of the shop shining a bright silver, with absolutely beautiful clean welds. Knowing that a human life was at stake made a difference.

Huntsville was a pretty sleepy town in those days. It had no airport, so we had to fly into Birmingham, frequently on an old Douglas DC-3, then drive north for two hours to Huntsville. The town was "dry," so if you wanted a glass of beer on a hot day you had to either get someone from the Army Redstone Arsenal to escort you into their officers' club or join the local Elks Club. Things changed enormously as the rocket team prospered.

Von Braun and his team went on to design many successful American missiles and launch vehicles, including the Jupiter-C–Juno, which put America's first payload into orbit; the Saturn 1, which launched Apollo development flights and supported the Skylab space station and made possible the Apollo-Soyuz handshake in orbit; and the Saturn 5, which sent the Apollo astronauts to the Moon.[6]

The Jupiter-C–Juno, a duplicate of which is in the Air and Space Museum, is a story in itself. In 1955 von Braun came to visit us at Rocketdyne, where I was at that time supervising preliminary design and analysis. The Redstone missile, powered by Rocketdyne's seventy-five-thousand-pounds-of-thrust engine and a direct descendant of the V-2, was flying reliably. Von Braun had calculated that with a small bundle of solid-propellant rockets fired in three stages on top of the Redstone, he could almost, but not quite, put a modest payload into Earth orbit. To achieve orbit he needed a 10 percent increase in the specific impulse (a measure of rocket efficiency) of our rocket engine. We put our chemists to work on the problem, and they came up with a mono-methyl-hydrazine additive to our ethyl alcohol fuel that would give the needed boost in performance and still act as an adequate coolant in the engine. Off von Braun went to Washington to get an expected approval.

Given the go-ahead, America could have been the first nation to put a payload into orbit around Earth, but when von Braun presented his proposal, President Eisenhower decided for a variety of reasons that he wanted any American spaceflight to be entirely a civilian endeavor with no military hardware involved.[7] Consistent with this policy, development was authorized for a modest new nonmilitary launch vehicle called Vanguard and its scientific payload under the leadership of the steady and personable Milt Rosen, an experienced engineer from the Naval Research Laboratory (NRL).

While Vanguard was still struggling through early development, the supposedly backward Soviet Union shocked and amazed the world by putting a relatively heavy spacecraft, Sputnik 1, into orbit on 4 October 1957, and quickly followed it with still heavier payloads into orbit. This achievement greatly boosted the image of the Soviets—export sales of their rather crude tractors and machinery suddenly boomed—and conversely was a real blow to American prestige, pride, and world-wide influence. Politicians rushed to voice their concern over America's lagging effort, and the "space race" was on. The next launch of Vanguard, on 6 December 1957, Vanguard's first attempt to put a scientific payload into orbit, was now given national television coverage and heaps of advance publicity. Unfortunately, it was another fiery, explosive failure, which had a devastating impact on national pride.

Eisenhower finally gave the go-ahead to von Braun, who had been pacing in frustration for two years. His Redstone rocket, modified with a spinning second- and third-stage package of solid rocket motors and relabeled the Jupiter-C, was ready, and von Braun's chief scientist, the gentlemanly scholar Ernst Stuhlinger, had planned a simple but meaningful payload. A physicist at the University of Iowa, James Van Allen, had developed a small cosmic-ray detector (essentially a Geiger counter to measure charged particles), which he had flown on sounding rockets and hoped to orbit on the Vanguard. In just ninety days William Pickering's crew at the Jet Propulsion Laboratory in Pasadena, California, integrated Van Allen's instrument with a fourth-stage solid-propellant motor, delivered it to Cape Canaveral, and integrated it onto the Jupiter-C in a configuration known as Juno-1. An army team (JPL was then an army lab, as was ABMA) put America's first payload, dubbed Explorer 1, into orbit on 31 January 1958. What made the

Left to right: William Pickering, James Van Allen, and Wernher von Braun in February 1958 holding up a model of Explorer 1, America's first payload into orbit and the discoverer of the intense Van Allen radiation belts surrounding Earth. (Courtesy Jet Propulsion Laboratory)

event truly historic was that Van Allen's little Geiger counter on Explorer 1, plus input from later Explorers, discovered the great radiation belts around Earth, now appropriately named the Van Allen belts, something the much larger Russian orbiters had failed to detect. The unlucky Vanguard finally made it into orbit soon afterward, on 17 March 1958.

Following the orbital achievement by an army team, Eisenhower got back to his "civilian" space program by creating the National Aeronautics and Space Administration on 1 October 1958. Based on the longstanding National Advisory Committee for Aeronautics (NACA), with its centers in Hampton, Virginia, Moffet Field in California, and Cleveland, Ohio, the new organization also integrated the people and facilities of JPL and von Braun's ABMA, along with space scientists from the NRL.

By the mid-1950s von Braun had become recognized as clearly the world's most effective spokesman for space exploration (later to be rivaled only by Carl Sagan). His fame as a salesman for space was firmly established in 1952, when he authored a series of beautifully illustrated articles in then-popular *Collier's* magazine proposing a very large Earth-orbiting space station featuring an immense spinning ring structure like the one in the later movie *2001: A Space Odyssey*. A follow-up article in April 1954, even more inspiring and again beautifully illustrated in full color by Chesley Bonestell, described in detail a mission carrying a human crew to the surface of Mars.

Before that time, if you were employed as an engineer as I was, you did not dare seriously discuss spaceflight. To do so would label you as a dreamer, not to be trusted with important calculations. (Who wants to drive across a bridge designed by an engineer whose head is in space?) After the *Collier's* articles the ARS invited von Braun to give a talk on his proposed projects to a technical audience at a meeting of their southern California chapter. The large Roger Young Auditorium in Los Angeles was packed with engineers geared up to shoot down this dreamer with a barrage of detailed technical questions. After all, most of us were inspired by the technical challenges, not by works of science fiction. To everyone's amazement, my own included, von Braun responded with well-thought-out answers to even the most detailed technical questions. It was a true turning point for space enthusiasts. From then on it was acceptable to talk seriously about spaceflight. I found out that most of the engineers I had been working with at Rocketdyne were there for the same reason I was: to tackle the difficult technical barriers and advance the new frontier of space exploration. We were designing missiles for the defense of our country, to be sure, but we were more excited about their potential as launch vehicles for spacecraft, and we were grateful to von Braun for bringing that goal out of the closet.

At MSFC von Braun was the undisputed leader of his team. They called him "the Professor," an honorary title given him by Hitler after the first successful flight of a V-2. The members of his team would faithfully follow his lead, even if they believed him to be wrong. In mid-1958, for example, after I joined von Braun and Herman Koelle, von Braun's chief preliminary design engineer, in successfully pitching the development of the Saturn I launch vehicle to the Advanced Research Projects Agency (ARPA), I pleaded in vain with Wernher to let us redesign the thrust structure that mounted the eight Rocketdyne H-I rocket engines to the Saturn I booster. I pointed out that we could easily save over two thousand pounds. He said, "No, I want it so simple you can analyze all the loads in just five minutes using only a slide rule." In keeping with that conservative approach, instead of requiring new large tanks, the propellants were carried in a bundle of existing Redstone and Jupiter tanks. As a result of this simple but heavy approach the Saturn I was soon put out of business when the much lighter Titan 3 became operational. By bundling existing propellant tanks together and using simple structures, however, von Braun got a Saturn I launch vehicle of impressive capability at an early date. When it flew in October 1961 it put the United States ahead of Soviet payload capability for the first time in the space race. It launched developmental missions for the Apollo Program and carried our astronaut crews to the Skylab space station and to the Apollo-Soyuz handshakes in space. And it flew with 100 percent reliability.

We at Rocketdyne managed to slip one major advancement in technology onto the Saturn I. In a cooperative effort between my Advanced Projects people and a bold test engineer, Paul Castenholz (who later was promoted to project manager of the all-important SSME, or space shuttle main engine), we started eliminating subsystems from the V-2–Redstone engine design and verifying the simplifications on an experimental engine labeled the X-I. The result was the elimination of complex pneumatic, hydraulic, and electrical subsystems of the V-2 engine, so that all that was required was two wires to start the X-I engine. Running current through the two wires ignited a squib in the bipropellant gas generator, which started the turbopump spinning. The building propellant pressure opened the main valves that let propellant into the thrust chamber, and propellants going back to the gas generator bootstrapped the engine up to full thrust. Beautifully simple. It went into production as the H-I engine, used in a cluster of eight for the Saturn I. It did pain me a bit that such an advanced engine should be used with such an ultraconservative launch vehicle.

In 1970 von Braun was talked into leaving his position as director of

MSFC and joining the team at NASA Headquarters struggling to define and sell a reusable launch vehicle, later to be called the space shuttle. At Headquarters von Braun no longer had his accomplished MSFC team to lead, and he clearly missed that, but he made a major contribution by finally getting the message across to Jim Fletcher, NASA's administrator, that a fully reusable launcher, with all elements flown back to the launch area, just could not be developed within any realistic NASA budget. It pained him to make this point, because he had been pushing reusability for many years, believing it to be a feature necessary to making his proposed space station practical. He knew, though, that development of a fully reusable shuttle would consume the entire NASA budget and wipe out all unmanned missions, which would be unacceptable to everyone. So the shuttle was compromised with a throwaway external tank and solid-propellant boosters that could be partially salvaged. It was a necessary step, even though the boosters would prove to be a future weak point.

A few years later, in 1977, five years after he departed NASA, I was talking with Wernher and asked him how he was feeling after recent surgery. He replied simply, "Not so good, Bob" and changed the subject. I knew he meant that he had received a bad report from his doctors. He died a few weeks later. The world lost a true giant in the advancement of space exploration. I am not a professional historian and am not qualified to debate with them details on whether von Braun was ever truly a Nazi and whether he approved of the SS's use of slave labor to build V-2s. I only know that I worked with the man and found him to be apolitical and totally dedicated to spaceflight. Like him, I designed rocket weapons for my country, but our focus was always on the future use of rockets for space exploration.

Rapid Advances in Launch Vehicles

Wernher von Braun was in a class by himself, but others made similarly major contributions to spaceflight. Consider the role of Belgian-born and MIT-educated Karel "Charlie" Bossart, a structures engineer at the Convair Division of General Dynamics in San Diego, California. Charlie had the wild idea that he could make a launch vehicle out of "balloons." A cylindrical toy balloon is made of very thin flimsy material, but when you blow it up it becomes quite stiff. So Charlie proposed building a vehicle with cylindrical propellant tanks made of tissue-paper-thin stainless steel that would not support even its own weight but would become rigid when pressurized. By his calculations such a rocket-powered missile would be so light that it could carry a nuclear warhead to the strategically required intercontinental range of fifty-five hundred nautical miles. He came in 1951 to Rocketdyne (then a small propulsion group within the Aerophysics Laboratory founded by Caltech professor Theodore von Kármán's pupil Bill Bollay at North American Aviation), asking for assistance in designing the propulsion system for his proposed intercontinental ballistic missile (ICBM). The idea was so "far out" that Rocketdyne would only initially assign a very junior engineer, yours truly, to work on the preliminary design, and I was already pretty busy working on the booster rocket engines for the Navaho intercontinental ramjet cruise missile.

Bossart's project, which he called Atlas, did sound unlikely. The Department of Defense (DoD) wanted any intercontinental missile to be able to carry a thermonuclear fusion warhead—the hydrogen bomb, or "H Bomb," that the Atomic Energy Commission (AEC) believed it could develop but whose weight was unknown. Using a conservative estimate of the hydrogen warhead weight, Bossart's ICBM would require a takeoff thrust of 1.2 million pounds, even using the most exotic of propellant combinations, liquid fluorine and ammonia. The largest engine we were working on then had a thrust of only 120,000 pounds, and more often than not it burned out after only about ten seconds of firing time. That was using the rather tame propellant combination of liquid oxygen and JP-4 jet fuel. Liquid fluorine was something else again. It attacks even glass, it will burn water, and it is lethal to breathe in concentrations as low as one part per million. Rough stuff. North American Aviation engineers had tested it in very small motors (at Los Angeles Airport under conditions that OSHA and EPA would never permit today) but could not imagine working with it in large quantities.

Several years later Rocketdyne built a special fluorine-safe test facility in the remote Santa Susana Mountains with an exhaust scrubber and a neutralizing pit to collect any spilled fluorine. During a test firing the fluorine ate through the gasket on its boilerplate storage tank. The pit caught the leakage, but a large brown cloud of gaseous fluorine formed. The crew in the pressurized blockhouse was safe, but the heavier-than-air cloud of fluorine did not dissipate and headed downhill, burning up weeds and cigarette butts as it moved along. As soon as they deemed it safe the test crew ran out of the blockhouse, hopped into a Jeep, and drove off to track the brown cloud as it headed toward the town of Canoga Park, a few miles downhill. Fortunately the cloud finally dissipated before reaching habitation. That experience did not encourage us on the practicality of full-scale launch operations with fluorine.

The estimates on hydrogen warhead weights kept coming down until Bossart's team figured they could get by with 840,000 pounds of take-off thrust, and this with our "tame" propellants, liquid oxygen and JP-4 (later reformulated and labeled RP-1 for rocket use). They would use seven of our Rocketdyne 120,000-pound engines, dropping off five part way through the boost phase to create a one-and-a-half-stage ballistic missile. The balloon tanks were so light it was not necessary to go to a more complex two-stage configuration. About this time a Soviet nuclear test was detected with a yield of a very large four hundred kilotons, boosted by the use of thermonuclear material, which caused great concern in Washington. On 14 May 1954 (just before I got married) we not only got the go ahead on Atlas but suddenly we were 1-A top priority, top secret, and, essentially, "money is no object."

Rocketdyne was required to build a major team in a hurry. Air force veteran Doug Hege, a capable and energetic manager who was not at all awed by the challenge of the project, was named to head the Atlas effort at Rocketdyne. Having been on Atlas from its beginning, I was quickly promoted to supervisor and then group leader for Preliminary Design and Analysis. I had to get not only top secret security clearance but also Q clearance from the AEC because I knew the projected warhead weight. We were all convinced that if the Soviets developed such an ICBM before we did that they would use it, either attacking the United States or else using it as a threat to expand their domain. That may sound unrealistically paranoid today, but in the days of Stalin and the table-pounding Khrushchev it was a realistic view. As I started into married life with my Anne, I was driving an hour each way to work and spending more than eighty hours a week on the project. Other than a quick kiss hello and goodbye I only saw my lovely green-eyed and red-haired bride on Sundays, and I was pretty tired then. Not a great way to start a marriage. (We subsequently had six beautiful children, so I guess we recovered, and then some.)

Atlas warhead weight estimates were reduced again, so we shifted to just four 120,000-pound booster engines and one sustainer engine for a total of 600,000 pounds of takeoff thrust. By that time our Rocketdyne team had already designed a cluster of three engines upgraded to 135,000 pounds each for a total of 405,000 pounds of thrust for the Navaho 3 booster, so 600,000 pounds for Atlas did not sound too wild. Then DoD and the AEC finally firmed up on a warhead weight that required only 360,000 pounds of thrust when using Bossart's balloon tanks. We uprated two booster engines to 150,000 pounds each and designed an entirely new high-performance sustainer engine with 60,000 pounds of thrust at sea level. That is the Atlas propulsion system that finally went into production.

Our Rocketdyne chief engineer, Paul Vogt, never believed those tissue-paper tanks were going to work. On a production line, before they were pressurized and filled out to shape, they looked just awful—like wrinkled prunes. So Paul was always leaning over my shoulder and encouraging me to double the design thrust, which he believed was going to be necessary later on when the tanks had to be beefed up. In that sense the first flight of an Atlas in 1957 was the most successful failure ever. The missile tumbled after reaching high speed, and the range safety officer had to hit the destruct button, resulting in a spectacular ball of fire overhead. Yet the sight of those balloon tanks staying rigid while tumbling end over end through the skies convinced all the doubters—even the ultraconservative Paul Vogt. Thus the United States developed a relatively small and highly efficient ICBM while the Soviets were forced to design much larger vehicles, a penalty that turned to their advantage later as we entered the space age.

Unlike the too-heavy Saturn 1, the Atlas was and still is a real hot rod, fully competitive with launch vehicles forty years younger, such as the European Ariane and Japanese H-2. In the 1960s and 1970s Atlases launched most of our robot spacecraft to the planets, and they continue to put up a multitude of geosynchronous communications satellites. As the father of the brilliant Atlas design, Charlie Bossart deserves a spot in our hall of giants.[8]

The USAF quickly recognized that it did not have the technical smarts in-house to properly direct the design and development of such an advanced machine as the Atlas ICBM, so it commissioned missile designers Sy Ramo and Dean Woolridge to assemble a team of technical experts. Accompanied by a secretary, the two men started out in a literal "little red schoolhouse" in Inglewood, California, where they quickly assembled a competent staff. The subsequent Ramo-Woolridge Corporation was later split into Thompson Ramo Woolridge (now TRW) and the Aerospace Corporation, which continues to provide technical expertise to the USAF.

Another outstanding performer during those hectic Atlas-development days was Al Donovan, who oversaw both the ICBMs (Atlas and Titan) and IRBMs (intermediate-range ballistic missiles; the Thor and Jupiter) for Ramo-Woolridge. Al never lost his cool and was a pillar of stability and good judgment. He subsequently became president of the Aerospace Corporation.

In 1954, when there were still some doubts about the Atlas' balloon tanks, the USAF decided to start a second-source ICBM program, called Titan, with Martin Aircraft as the prime contractor and with Aerojet Corporation developing the rocket engines. Rocketdyne was supposed to be a backup to Aerojet, so I spent considerable time with engineers at both Martin and Aerojet. Aerojet, one of the pioneers in rocket development, grew out of Prof. Theodore von Kármán's early work with Frank Malina at Caltech. Although the Aerojet engineers and chemists were experts and leaders in solid rockets, they had never built any very large liquid-propellant engines. Those they had tested and developed used storable propellants such as nitric acid rather than cryogenics such as liquid oxygen, which they were now being asked to use.

In 1955, when the ARS somehow managed to arrange a classified tour of Rocketdyne's Santa Susana test facility, the Aerojet people understandably rushed to sign up. As the tour group assembled for the evening tour, tall and good-looking Bill Cecka, head of the facility, announced that they were first going to witness a full-duration test of the Thor engine, then the Atlas booster engine pair, then the Atlas sustainer engine, and finally the full 405,000-pounds-of-thrust Navaho 3 booster engine assembly, by far the most powerful in the free world. Bill then casually mentioned that he had to leave to catch an airplane but he was sure everything would go on schedule. I cringed, because in those days we still blew up engines more often than we liked, and there were always lots of little anomalies that could delay tests for several hours.

I was the tour guide on one of the buses, and there in the front row of the bus was Bob Young, manager of Aerojet's Liquid Rocket Plant. The person seated next to him I believe was his chief engineer, R. C. Stiff. The first test, the Thor engine, went off right on time and to full duration. In those days we stood out in the open, fairly close to the test stand. Very impressive—you did not just hear the roar, you felt the power shaking your chest and hurting your ears (we never thought of ear protectors, so I now have no hearing above 2,000 Hertz). Back on the tour bus there was excited chatter. Next was the 300,000-pound fury of the two Atlas booster engines, right on schedule. Back on the bus, lots of excitement. Then the Atlas sustainer engine, with its beautiful translucent exhaust from a high-expansion nozzle, again right on schedule and for the planned duration. In the now-noisy bus I looked at Bob Young. He and his chief engineer had their heads down and weren't talking at all; they realized that Aerojet had a long way to go to catch up with the crew at Rocketdyne. Finally, the full power of the 405,000-pound-thrust Navaho booster went off within minutes of schedule. I tried to act like this was just another routine evening at Santa Susana, but I was probably more impressed—especially with the confident nerve of Bill Cecka—than anyone on the bus. Rocket engine development in the United States had certainly come a long way from the pioneering days of Robert Goddard.

That 405,000-pound thrust of the Navaho 3 cluster was indeed impressive—it would shake you until your teeth rattled—but the march to greater rocket power did not stop there. In about 1956 the USAF propulsion people at Wright Patterson Field in Dayton, Ohio, asked Rocketdyne what the upper limit of thrust was for a single rocket engine. We decided to tackle the question by picking the highest thrust level any of us could imagine and then trying to design it in detail, complete with stress analyses. The highest figure any of us, including the senior designer in our group, Jack Conyers, could imagine for a single thrust chamber was 1 million pounds, so we started a design with that level in mind. Some of the required performance figures were awesome. For example, it required forty thousand horsepower just to drive the turbopump, which had a propellant flow rate high enough to drain an Olympic-sized swimming pool in about ten seconds. And yet analyses indicated that there appeared to be no structural reason why the engine could not be built. The USAF people said OK, go ahead and see if you can build a prototype. The result was that by the time NASA was formed in 1958 Rocketdyne had a thrust chamber firing regularly at 1 million pounds of thrust and its accompanying turbopump well along in construction. It was a relatively easy step to uprate it to 1.5 million pounds, at which level it was named the F-1 and mounted in a cluster of five to power the Saturn 5 launch vehicle that sent the Apollo astronauts to the Moon.

A number of people at Rocketdyne made important contributions to spaceflight. Bill Bollay organized the parent Aerophysics Laboratory at North American Aviation and hired Sam Hoffman to assemble a team of rocket engineers. I was most fortunate that my first boss there was George P. Sutton, author of the first comprehensive and still-popular textbook on rocket propulsion design and one of the early presidents of the American Rocket Society. George taught me a great deal about propulsion theory as well as practical design. Best of all, he was willing to delegate responsibilities to me just as fast as I was willing to take them on. At first I considered George a "cold fish," very serious and formal. The death of his young wife in a terrible traffic accident (a truck ran a red light and hit her car broadside) probably contributed to his sober demeanor, but he warmed up with time, and I soon realized he was really interested in his subordinates as people, not just workers. We got along beautifully, even under the pressure of those eighty-hour weeks in a hot, dusty barn of a temporary facility in East Los Angeles, and still do today. George is semiretired now but teaches regular courses in rocket design to classes of government and contractor employees all over the United States.

A good measure of the credit for Rocketdyne's success, including development of the F-1 and J-2 engines for Apollo's Saturn 5 and the SSME engines that power the space shuttle, should go to two men, each outstanding in their unique contributions: Tom Dixon, Rocketdyne's vice president, and Paul Vogt, chief engineer. Tom, the number two man under Sam Hoffman, was the enthusiast, quick to grasp new concepts, to see ways to advance rocket capabilities, and bold and decisive in making things happen. Tom was a marvel to me. He did not fit any textbook description of a successful executive—he wasn't especially impressive looking, nor was he a great speaker or facile with words, but he was 100 percent sincere and dedicated to advancing the rocket capability of his country.

When Tom decided that something needed to be done you knew that his judgment was sound and focused on the national good, not just politics or corporate profits. In 1957, when I presented him with some study results that pointed firmly to liquid oxygen and liquid hydrogen as the desired propellant combination for many future uses, even though liquid hydrogen seemed too exotic at the time, Tom rushed me off to the Pentagon to give my briefing to Defense Department officials. Without appointments he was able to get us into all offices, including the secretary of defense. When they heard it was Tom Dixon in the office they knew he would not ask for time unless he believed it important for the country, and they made time available in their overbooked day. Amazing.

When we got back to the plant in Canoga Park, Tom did not want to wait for new government funding to arrive. He assembled my team and detailed design and test engineers, such as the energetic and results-oriented Paul Castenholz, and told us, "For the good of the country *you* [Tom never said *I*] must build a hydrogen-oxygen engine of about 200,000 pounds' thrust and get it running as soon as possible." He brushed aside questions about funding and liability in case we blew up one of the test stands, which legally belonged to the government, and said simply, "I know *you* can do it." So we did. And when NASA needed a high-performance hydrogen-oxygen engine for the second and third stages of the Saturn 5, the Rocketdyne J-2 engine was practically ready.

Paul Vogt, a large man with a piercing gaze, was the complete complement to Tom Dixon. Very conservative, Paul had the capacity to sit for hours analyzing minute details such as the design of a single bolt, but for new ideas he was the kiss of death. He was very clever in this: first opposing your new concept, so that you would sell it harder and harder, then suddenly on your side, so enthusiastic that you found yourself trying to hold him back from going overboard by pointing out the idea's limitations. Then, wham, he took all your identified limitations and threw them back at you. I learned to always wait for Tom Dixon to get back in town before proposing anything new. Nevertheless, it was Paul Vogt's dogged pursuit of fine details that took Rocketdyne's big engines from blowing up after a few seconds of firing to where our astronauts could climb aboard their launch vehicles with confidence. Unfortunately, Paul Vogt had a bad stroke a few years ago and died recently. Tom Dixon was engaged in a wide variety of activities until 1997, when doctors detected widespread cancer, and he passed away in early 1998. Tom was a lovely person whose vitality and enthusiasm were infectious. He will be missed as a pioneering rocketeer but even more so as a friend.

Before leaving the subject of launch vehicles I want to pay tribute to another pioneer in the field, Krafft Ehricke. Krafft was a junior member of Wernher von Braun's team in Germany. When he came to America after the war he went with General Dornberger to work at Bell Aircraft. A prolific worker, I do not know when Krafft found the time to sleep. At a symposium I would have dinner with Krafft and he would rush away to work up more graphs for a talk he was to give the next morning. You always braced yourself for one of his talks; he would race enthusiastically through complicated graph after graph of fascinating computations while the session chairman kept signaling to no avail that his allotted time was up. Still, you were sure to go away with lots of new knowledge about spaceflight.

In 1954 Krafft joined Convair in San Diego and very shortly came up with the concept of a high-performance hydrogen-oxygen upper stage for the Convair Atlas to be called Centaur. He came to us at Rocketdyne for expertise in designing the propulsion system, and I then joined Krafft and Convair's Frank Dore in briefings that sold the concept to the USAF and the Advanced Research Projects Agency. We at Rocketdyne then awaited the usual competition for the engine contract, which we were certain to win. It was the shock of my early career when Col. Norman Appold of the USAF, assigned to ARPA, announced that he was awarding the engine contract sole source to Pratt and Whitney (P&W). Pratt and Whitney had never built any kind of rocket engine before. I still do not know how Appold got away with that. Today there would be a huge protest, but in those days you did not argue with your customer. Later, when I was told that Appold after retiring from the USAF had gone to work for P&W's parent corporation, I thought, Aha, now there's bound to be a congressional investigation. Sure enough, there was, but it was concluded that all was legal and proper. In truth, I must grudgingly admit that P&W had some prior classified experience working with liquid hydrogen for an advanced turbojet engine, and it subsequently did a fine job with its new RL-10 rocket engine for the Centaur (in fact, RL-10s continue to perform well today). Krafft Ehricke's Centaur stage has been a terrific thoroughbred, too, delivering spacecraft all over the solar system atop both Atlas and Titan boosters and remaining to this day a mainstay in sending communications satellites into geosynchronous orbit.

By 25 May 1961, when newly elected president John F. Kennedy initiated the extraordinarily bold Apollo Program to land humans on the Moon, rocket propulsion and launch vehicles had advanced to the point where they were no longer the restraint to space exploration. At Rocketdyne the Atlas, Thor, and Jupiter engines were all running reliably, and we even had early models of the F-1 and J-2 engines in development well in advance of their future application on the Apollo Program's huge Saturn 5 launch vehicle. We had greatly simplified engine design with an experimental X-1 engine that was put into production as the H-1 for the Saturn 1 and then later called the RS-27 as it powered the highly reliable Delta launch vehicles. My group had patented and done advanced development on a very high performance engine configuration called the Aerospike Engine, which Rocketdyne,

The Atlas-Centaur launch vehicle. Initially with an
Agena upper stage and later with a Centaur high-
performance upper stage, the lightweight Atlas sent
Mariner and Pioneer spacecraft on their way to Venus,
Mars, Mercury, Jupiter, and Saturn. (Courtesy NASA)

under the leadership of my very talented successor as Rocketdyne's
chief of advanced projects, Sam Iacobellis, later proposed for the
shuttle but which was deemed to be too advanced. (In 1996, forty years
after its conception, a linear version of the Aerospike Engine was selected to power NASA's X-33 hypersonic reusable launch vehicle.) All
in all, rocket propulsion and launch vehicles were in pretty good shape.

The greater challenge to space exploration was then to develop reliable spacecraft to probe the wonders of our solar system. In this field, technical progress was coming very grudgingly. That was where the technology frontier was now most challenging, so that is where I wanted to work, and I began to look for an opportunity to contribute.

Struggles with Early Spacecraft

With funding from ARPA and NASA in 1958 and 1959, four Pioneer spacecraft—three unsuccessful, one a partial success—were launched to fly by the Moon.[9] When NASA was created in October 1958 to take over space exploration projects, the Ranger Lunar Program was started as a serious venture into deep space. The program goal was to land a scientific payload on the surface of the Moon. The Jet Propulsion Laboratory was transferred from the army to NASA and was given project responsibility for Ranger—the beginning of what would become JPL's long and historic role in planetary exploration.

The origins of JPL go back to the 1930s, when the legendary Prof. Theodore von Kármán at Caltech and a group of his graduate students, headed by Frank Malina, were experimenting with small rocket motors at a then-remote site in the Arroyo Seco gully, just north of Pasadena's Rose Bowl. During World War II they concentrated on the successful development of solid-propellant rocket motors for aircraft jet-assisted takeoff (JATO). In order to continue this development von Kármán organized the Aerojet Engineering Corporation and, in 1944, the Jet Propulsion Laboratory, with Frank Malina as the first director of JPL.[10]

By the early 1950s Louie Dunn had succeeded Malina as JPL's director and was continuing the lab under the sponsorship of the U.S. Army as a low-key research facility for aeronautics and rocket propulsion technology. At Rocketdyne during that period, we were trying to advance from heavy-walled thrust chambers of the V-2 type to much lighter thrust chambers made up of a bundle of thin-walled tubes through which the rocket fuel would pass as coolant before being combusted. We knew that the fuel would boil as it passed through these tubes and that the trick was to restrain the process to "nucleate boiling," with many small bubbles, rather than "film boiling," which would completely blanket the tube walls and cause them to burn through. I found that the only source of appropriate nucleate boiling data was at JPL, but the lab in those days took a leisurely approach to research, and its test results might not be published for another five or six years. To get the data I needed I periodically visited Fred Gunther and his heat transfer lab at JPL, drank coffee and chewed the fat about skiing, camping, and sailing for an hour or so, and then casually asked if I could browse through the lab's file of test results on nucleate boiling. Thus did JPL make a little-known contribution to the design of the modern generation of liquid-cooled thrust chambers.

In 1951 Dr. William Pickering, a low-key but strong-willed New Zealander of Scottish descent, succeeded Dunn as JPL's director, and under his leadership the lab began to transition from just applied research into flight-system development. Research effort in subsonic and supersonic wind tunnels and into combustion processes gradually disappeared in favor of the development of operational rocket systems and spacecraft, and JPL had to gear up from its relaxed research pace to meeting schedule milestones. The solid-propellant Sergeant missile and the liquid-propellant Corporal missile were major development projects successfully completed, and Explorer 1 was an exciting ninety-day rush project that was followed by a pair of mainly unsuccessful lunar Pioneers. Pickering could see that space was the new frontier and wanted JPL to redirect its energies from missiles to lunar and planetary exploration. He successfully transferred the lab from army to NASA soon after the latter's creation. Effective 1 January 1959, JPL became a NASA center, and its first major venture into space was the NASA-funded Ranger Project, which proved to be a real struggle.

Rangers 1 and 2 were developmental flights intended to travel well past the Moon. Rangers 3, 4, and 5 were more ambitious craft incorporating a complete lunar landing system, including approach radar and a solid-propellant retro motor to kill the eight-thousand-miles-an-hour lunar approach velocity. If the system worked, it would place its scientific payload, a highly sensitive seismometer, on the surface of the

Moon in order to measure moonquakes and provide clues to the interior structure of the Moon. By 1961, however, the Ranger Project had yet to achieve any success.

Then, on 25 May 1961, just a few weeks after Yuri Gagarin became the first man in orbit (on 12 April), President John F. Kennedy made his historic speech to Congress in which he dedicated the nation to landing Americans on the Moon and returning them safely to Earth before the end of the decade—an enormous technical challenge that, if successful, would clearly demonstrate American technical superiority over the rapidly advancing Soviet Union (and some say also make the public forget Kennedy's embarrassing Bay of Pigs experience). It still astounds me that any president would be so bold and daring and would also be able to sell such a pioneering project to the nation. His pro-space and former Senate-leading vice president, Lyndon Johnson, was a big help in getting full support from Congress for this superambitious Apollo Project.

Little was known about the surface of the Moon at that time. A respected scientist, Dr. Thomas Gold of Cornell University, had conjectured that after billions of years of asteroid bombardment the surface of the Moon would have been pulverized into a fine powder. On a microscopic scale these micron-sized particles of powder would be quite jagged and tend to link together. Under the Moon's low gravity, and with no atmosphere to disturb them, Tommy Gold believed they would build into what he called "a fairy castle structure," which would have an extremely low bearing strength. Moreover, he believed this fluffy dust layer could be up to ten meters (over thirty feet) deep. Apollo landers would sink completely out of sight into this lunar quicksand, which Tommy Gold's dubious fellow scientists immediately labeled "Gold Dust." The known thermal properties of the Moon did not seem to support the existence of an insulating surface layer of dust, but the NASA Apollo planners could not ignore the possibility, however slight, of a landing catastrophe.

I was convinced that the Ranger lunar lander could do a lot more to aid the Apollo lander design than just deliver its chosen seismometer instrument. It should be able to measure the bearing strength of the lunar surface and determine whether Gold Dust was a real threat. The contractor for Ranger's lunar landing system, the Aeronutronic Division of Ford Motor Company, located in Newport Beach, California, was gracious enough to hire me in the fall of 1961 and soon after made me its chief engineer for Space Systems.[11] The company's beautiful new facility was situated on a hill overlooking the blue Pacific, Santa Catalina Island, and Newport Harbor and was like a country club compared to no-nonsense Rocketdyne. Aeronutronic was the brainchild of Henry Ford II, who was proud of the production by the Ford Motor Company of equipment to help win World War II and wanted Ford to continue to contribute to defense efforts. Of course, he also believed that would give him more leverage with whomever happened to be president of the country. He had hired some top talent, so that when I arrived I found a lot more chiefs than Indians. It seemed like everyone was a recognized expert: I would poke my head into an office to introduce myself and see several authoritative textbooks on the shelf authored by the occupant. I also found that no one knew who my boss was supposed to be; I was on my own.

Now, this requires a bit of explanation. Aeronutronic was entirely different from my former employer, North American Aviation, and not just in the relative elegance of its facilities. North American was run by engineers—they went for technically challenging projects, such as the hypersonic rocket-powered X-15 airplane, the Apollo spacecraft, and the even later space shuttle—and didn't really worry all that much about whether the project would be profitable. At the Ford Motor Company, though, profit was all-important. They wanted to build their aerospace business rapidly at their Aeronutronic Division, but they always weighed the profit potential, and profit to them meant long production runs. This was going to make it tough for me to convince them to bid on space exploration projects, which usually called for only one or two spacecraft and had no long-term production potential.

However, there was another unique feature of Aeronutronic that worked in my favor. Ford traditionally developed its future plant general managers by rotating them into varied management positions. They would take a promising financial manager and place him for a year as head of engineering, then a year as head of manufacturing, and so on. It was like musical chairs every year at the top management level.

Rather chaotic, but effective for management training. They hired me without any specific position in mind just because they thought I would pursue new business prospects, and it was not that unusual that, having arrived right after another round of musical chairs, no one really knew who my boss was supposed to be. This was perfectly fine with me, and I immediately took off on my own and started preparing a proposal for Ranger follow-on missions in preparation for Apollo.

I found that Aeronutronic engineers under the direction of Frank Denison, Bill Hostetler, and Bob Neese had already designed a clever little line-scanning camera that would permit future Rangers to take pictures on the lunar surface. It was simplicity personified, just a light meter that nodded vertically while slowly rotating. There was no wind on the airless Moon so nothing moved—even shadows moved very slowly during a lunar day, which lasted almost one Earth month—so a high-resolution picture could be returned to Earth at a very leisurely low-power bit rate. Yet this was not going to work if the Moon's surface were indeed covered with Tommy Gold's deep Gold Dust. Not everyone, or even many, believed in Tommy's theory, but it was possible[12] and therefore potentially disastrous for an Apollo landing on the Moon. So just three weeks after reporting for work at Aeronutronic, and without any top management review or approval, I submitted a formal proposal to JPL for a Ranger surface measuring capsule (SURMEC) that would deploy an array of small baseball-sized penetrometers to sample lunar-surface bearing strength.

The Ranger project manager at JPL was a talented and dynamic engineer named James Burke. Today, more than thirty-five years later, Jim still has his athletic good looks and still exudes energy and enthusiasm. He was excited about our proposals for a lunar facsimile camera and surface measuring capsules for Ranger and helped sell them to NASA Headquarters. Within a month we had three signed contracts with JPL to begin key parts of the development. Progress was so promising that in early December 1963 we received a phone call from JPL saying they were sending their helicopter down to pick us up and fly us to JPL to sign a contract for an additional twenty Ranger capsules and systems to land them on the Moon. Wow, twenty deliverable systems! Here was "production" for the Ford financial people. We were excited. Our excitement took a startling twist, though, because we never made it to JPL that day. About half way from Newport Beach to JPL in Pasadena our chopper experienced a catastrophic failure. We crashed in Rose Hills Cemetery in the Whittier foothills. Miraculously, the pilot and we three passengers managed to crawl away without serious injury. We declined JPL's offer to send another helicopter to pick us up—the next day we drove by Ford automobile with its four wheels planted firmly on Mother Earth to sign our new contract.

Unfortunately, although the Ranger lander and seismometer capsule effort at Aeronutronic was making progress, JPL's in-house effort to develop the Ranger bus spacecraft to deliver the capsule systems on a course to the Moon was not going well. Oran Nicks's *Far Travelers: The Exploring Machines* offers a good summary of the difficulties encountered. First of all, designing a craft to travel into outer space was a new venture; there was no successful prior experience to build on. Second, the project was being undertaken on a minimal budget. Third, the Ranger Program was considered an experimental project to develop technology for sending future payloads to Mars; with this in mind it was decreed that the entire Ranger vehicle had to be sterilized before launch, a complicating procedure unnecessary for the Moon but required in the future to prevent biological contamination of Mars. That meant that everything on the Ranger spacecraft had to be baked at high temperatures or else assembled in a glove box containing a toxic and corrosive ethylene oxide atmosphere. This was no way to develop and build reliable electronics and mechanisms. A fourth problem was that the Atlas-Agena launch vehicle was still not reliable.

As a result, Rangers 1 and 2, launched in 1961, never made it beyond low Earth orbit instead of flying past the Moon. In 1962 Ranger 3, the first lander configuration, missed the Moon. Ranger 4 suffered a bus computer failure and collided with the back side of the Moon, and Ranger 5 had a bus power failure and also missed the Moon. These were tough times for the fledgling U.S. civilian space program. At the time the Soviets were launching many successful missions into Earth orbit and to the Moon, and there was great pressure for some American space achievements. After the five Ranger failures congressional committees demanded major reviews of the entire project. When a review team headed by Cmdr. Albert Kelly of the U.S. Navy came to Aeronutronic they chewed me out for not doing more all-up system testing of the Ranger lunar landing system. I told them I agreed completely with the need for the tests, but that it was just not possible to conduct them within the small ($5 million) budget allocated to us.

The pressure was so great that there were serious proposals to shut down JPL and transfer its space work elsewhere. The lab looked even worse when it hastily modified a Ranger spacecraft, called it Mariner 1, and attempted to send it to the planet Venus. The range safety officer at Cape Canaveral blew it up as it strayed off course shortly after liftoff. A second attempt, with Mariner 2, was made three weeks later. This flight appeared to be doomed time after time. It seemed that everything went wrong. First the Atlas-Agena launch vehicle started into an unprogramed roll, finally spinning at the rapid rate of one full revolution per second. Then, for no known reason, it stopped rolling just before burnout and, miraculously, at a roll angle position less than two degrees off the intended value. It is no wonder that Wernher von Braun later commented that Mariner 2 must have used neither the conventional radio guidance nor inertial guidance but rather "divine guidance." Right after the salvaged launch the spacecraft's Earth sensor appeared to lock onto the Moon instead of Earth. After that was resolved, one of the two solar panels kept shorting out; by the time it died for good, the spacecraft was luckily close enough to the Sun that the one remaining panel could generate enough power to keep the spacecraft alive. Although by then internal temperatures were going off the scale, somehow the overheated Mariner 2 kept functioning. Its propellant tank was overpressurizing, but that seemed less critical than the overheating. On approaching Venus the spacecraft computer failed to turn on the instruments, but a backup signal from Earth arrived just in time for the instruments to make near-perfect scans across the planet. It may have taken tons of luck, but Mariner 2 was the first successful mission to another planet, and its success took a lot of political pressure off JPL.

Just a few days after the Venus encounter the overcooked spacecraft began to die, and in three weeks it was silent forever. Jack James, the Mariner 2 project manager at JPL, deserves a great deal of credit for his leadership of the team that nursed the perpetually ailing spacecraft

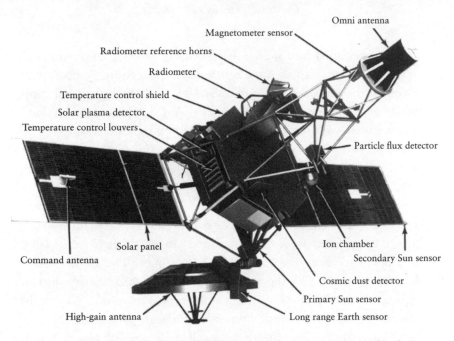

Omni antenna

Magnetometer sensor

Radiometer reference horns

Radiometer

Temperature control shield

Solar plasma detector

Temperature control louvers

Particle flux detector

Command antenna

Solar panel

Ion chamber

Secondary Sun sensor

Cosmic dust detector

Primary Sun sensor

High-gain antenna

Long range Earth sensor

Mariner 2, built by JPL and successfully launched to Venus in 1962 for the world's first successful close flyby of another planet. (Courtesy Jet Propulsion Laboratory)

all the way to Venus. It is not surprising that Jack, an impressively tall Texan who always seems alert and quick to grasp problems, went on to other Mariner successes.

The miraculous flight of Mariner 2 finally scored a big win for the United States in the space race (the Russians had made several unsuccessful Venus attempts beginning in 1960) and literally saved JPL. With the pressure eased a little, JPL and NASA reorganized the Ranger Project. Needing a scapegoat, I suppose, JPL removed Jim Burke as Ranger project manager. (Burke never did serve again as a project manager but concentrated on advanced concepts, where he had proven his valuable talent. After retiring from JPL he continues to be very productive as technical editor for the *Planetary Report,* an outstanding publication of the Planetary Society.)

In Jim's place JPL appointed Harris "Bud" Schurmeier, a more experienced and conservative engineer who was chief of JPL's Systems Di-

President John F. Kennedy (*far right*) at the White House congratulating (*left to right*) Jack James, Robert Seamans, Robert Parks, William Pickering, and James Webb on the success of Mariner 2, which, on 14 December 1962, became the first spacecraft to successfully reach another planet. (Photograph by Cecil W. Stoughton; courtesy NASA)

vision. If one had to pick one word to describe Bud it would be "solid." He looks solid and muscular, but more than that, he is unflappable and his judgments are solid. (Later on, when we were trying to sell the Grand Tour missions, I was going to insist that Bud be the designated project manager—I needed someone whose judgment could not be challenged.) NASA Headquarters dropped the lunar landing objective for Ranger, and the mission was scaled back to just trying to return television pictures of the lunar surface before the spacecraft impacted.

When the landing missions for Ranger were abruptly canceled on 13 December 1963 it came as a complete shock to the Ranger team at Aeronutronic. We knew that NASA's Ranger, Surveyor, and Centaur Programs were all facing budget overruns, but there had been no warning out of NASA Headquarters that Oran Nicks, director of Lunar and Planetary Programs at Headquarters, was even thinking about cancellation of the Ranger landers. In fact, even Nicks's own Ranger program manager, Newton W. "Bill" Cunningham, was so stunned that he called us immediately at Aeronutronic before more properly breaking the word to JPL. I was personally devastated—it is still probably the biggest disappointment of my space career. Our lunar facsimile camera at Aeronutronic was working flawlessly and producing clear, high-resolution pictures. I wanted us to take the first picture on the surface of the Moon so badly I could taste it. As some consolation, at least my lunar surface measuring capsules were adapted to be deployed from the Apollo lunar landing module on its approach to the Moon's surface as an aid to picking the best touchdown site and avoiding any hazardous Gold Dust areas.

After the first five failures Ranger was dubbed the "Murphy's Law Spacecraft" by Oran Nicks. Ranger 6, the first of the simplified television Rangers, continued the Ranger reputation. It was headed perfectly for the Moon, but the camera never came on, probably due to an arced switch. Walt Downhauer, a young and personable systems engineer at JPL, had been picked to narrate the lunar encounter for closed-circuit viewers at JPL. As the seconds to lunar impact ticked away he had to keep reporting no pictures. Finally the signal was lost at impact. He found it so unnerving that he refused to narrate any more spacecraft encounters. That role was taken over by a quick-thinking and smooth-talking JPL scientist, Al Hibbs, who came to be acknowledged as the "Voice of Mariner," as well as Viking and Voyager, and became a "professional" performer with his own science series on public television.

Finally, Rangers 7, 8, and 9 were complete successes, returning images of the lunar surface with a resolution at least a thousand times finer than the very best images from Earth-based telescopes. Not only was JPL saved, but a new capability for lunar and planetary exploration had been established. This was due to the efforts of many at JPL, but also to the foresight, judgment, management skill, and sheer determination of two men at NASA Headquarters, Edgar Cortright and Oran Nicks.

Cortright was an engineer at the Lewis Research Center (LeRC) of

NACA when it became a part of the new NASA. His former NACA boss, Abe Silverstein, brought Ed to Headquarters to be responsible for NASA's unmanned spacecraft development. I would personally vote Ed the most talented manager I ever knew at NASA Headquarters. He was of average build, not imposing in appearance like a James Webb or a Wernher von Braun, but his foresight and judgment were outstanding, he was marvelous in dealing with people, and his communication skills were exceptional. No matter the occasion, formal or otherwise, Ed always knew just the right thing to say. He could work more effectively with key congressional committees than anyone else I know. I always hoped he would eventually be named NASA administrator, but after a notable career as a civil servant he apparently felt he had to go off to industry to make enough money for a more comfortable retirement for himself and his lovely wife, Beverly. Who could blame him?

In 1960, to manage Lunar and Planetary Programs, Ed had hired Oran Nicks, who had previously worked as a supersonic aerodynamicist at North American Aviation and as a project engineer at Chance Vought. In addition to being a real gentleman, Oran was an engineer of excellent technical ability, sound judgment, high personal integrity, and exceptional perseverance. He had some of the speed of the hare but, more important, all of the determination of the tortoise, an all-important trait in the 1960s, necessary to create the technical foundation for planetary exploration. In 1963, when NASA found itself in serious trouble with the development programs for Ranger, Surveyor, and Centaur and could not sustain full funding for all three, it was Oran who argued for simplifying Ranger rather than canceling it outright. His perseverance led to eventual success with all three programs. However, in 1964, immediately after the rescue of Ranger, the Surveyor program took over the spotlight as a continuing challenge.

Surveyor was an ambitious step beyond Ranger. Ranger was a "hard lander," designed to survive a lunar landing impact of up to two hundred miles per hour. Surveyor, on the other hand, was designed to touch down softly on its three legs, just as Apollo would do a few years later. Because JPL was busy building the Ranger bus in-house, the task of designing and developing Surveyor was contracted to Hughes Aircraft. This was JPL's first experience working with a systems contractor. They assigned capable Walker E. "Gene" Giberson as their project manager but only gave him a handful of people to keep a watch over Hughes' shoulder.

Surveyor quickly got into deep trouble. Landing tests out in the Mojave Desert in 1964 were supposed to demonstrate the complete landing system; both test vehicles crashed due to multiple failures. Gene Giberson was abruptly replaced as project manager by his boss, Robert J. Parks, a well-liked and highly respected veteran of JPL's Sergeant missile development for the army and subsequently manager of Lunar and Planetary Programs at JPL. Giberson was reduced to deputy project manager but performed well and without resentment in his new assignment (later he was given another chance as project manager; see chapter 5). Parks was immediately given an additional two hundred people to work with the Hughes engineers. The JPL contributors eventually reached a total of six hundred.

Planetary Mission Capability Demonstrated

In 1966 Surveyor 1 headed for the Moon and, to the surprise of many, made a successful soft landing, returning television pictures that showed a rock-strewn dusty surface, but one that could readily bear the weight of the spacecraft. So much for Gold Dust. For me it meant the cancellation of the Apollo landing-aid system that we had successfully developed from my proposed Ranger SURMEC system. No big blow here, as I never thought it too probable that Gold Dust existed, and besides, we had developed some valuable new technology that could be applied to future penetrators to probe the surface of planets, asteroids, and comets.

Five of the seven Surveyors were successful, and Oran Nick's lunar program was on a roll. The subsequent Lunar Orbiter Program went much more smoothly, with the personable and capable Lee Scherer as program manager at NASA Headquarters, Cliff Nelson as project manager at the Langley Research Center, and the Boeing Company as system contractor. All five Lunar Orbiter flights were successful and returned beautiful pictures of more than 99 percent of the lunar surface.

Oblique view of the Moon's two-mile-deep crater Copernicus. Photograph taken by Lunar Orbiter 2 on 23 November 1966. The first released of the dramatic oblique views taken by a Lunar Orbiter, this photograph created a sensation when first shown by Robert Seamans to a large AIAA meeting in Boston in December 1966. (Courtesy NASA)

I recall being at a large meeting of the American Institute of Aeronautics and Astronautics (AIAA)[13] in Boston when Bob Seamans, deputy administrator of NASA, unexpectedly projected onto the auditorium's large overhead screen the first oblique view of the Moon from Lunar Orbiter. It was a spectacular panoramic photo of the crater Copernicus, showing steep cliffs around the crater rim and an unearthly surface of hills and boulders. There was a gasp from the crowd and then everyone spontaneously jumped to their feet and gave a standing ovation that lasted for minutes. Truly a tribute to the progress of NASA's Lunar and Planetary Programs.

Meanwhile, Jack James and his Mariner team had prepared two Mariner spacecraft for launch to Mars in 1964. To accommodate the reduced power of the Sun at Mars distance, they had augmented the solar panels from two (on Mariner 2) to four. Unfortunately, a new shroud made of composite honeycomb distorted after launch and trapped Mariner 3 in its launch vehicle. In the remaining precious days in the launch window, however, JPL and Lockheed engineers were able to design and fabricate a replacement all-metal shroud that allowed Mariner 4 to get off to the first successful flyby of Mars. Mariner 4 took twenty-one snapshots, strung across the Martian surface and of good enough resolution to show a very ancient looking, heavily cratered Moon-like surface.

This was a disappointing surprise to almost everyone, as astronomers, beginning with Giovanni Schiaparelli in 1877 and Percival Lowell in 1894, had reported a network of "canals" on Mars, as well as dark areas and surface color changes during the Martian seasons. By 1964 astronomers had found that the atmosphere of Mars was quite thin and that a low surface pressure would not permit liquid water, much less canals full of water. Still, it was thought that even if they were not canals, the linear features might indicate ridges or rifts—and something was causing the color changes. Overall, telescope observations certainly suggested a dynamic planet, not the ancient, Moon-like dead one being revealed by Mariner 4. We'll go into this more in later chapters on Mars exploration with the Mariner 9 and Viking spacecraft.

With the completed flight of Mariner 4, confidence in the success of planetary missions began to build. In 1967 Mariner 5, under the project leadership of Dan Schneiderman (who had been Jack James's deputy for Mariners 3 and 4), was another success, reaching Venus in time to disprove a Russian claim that they had reached the surface. Whereas the signal from the Russian probe had stopped at a pressure of twenty-five Earth atmospheres, Mariner 5 had convincing data that the Venusian surface pressure was a crushing ninety Earth atmospheres.

In 1969, under the leadership of JPL's Bud Schurmeier as project manager, Mariners 6 and 7 were launched on their way to Mars flybys. Design and development of these spacecraft had not gone all that smoothly, as JPL was attempting to advance the state of the art in spacecraft design. For example, the telecommunications bit rate was advanced from the snail's pace $8^{1}/_{3}$ bits per second of Mariner 4 to

16,200 bits per second. This permitted flying much more advanced science instruments on the new spacecraft. Heavy responsibility for the development of these instruments was given to principal investigators (PIs) at various universities. One such PI, Dr. George Pimentel of the University of California at Berkeley, was a recognized expert in spectroscopy but nevertheless a pain to deal with. This was "hippy" generation and student protest time at universities, Berkeley especially, and George seemed to be trying to be one of the gang with his graduate students by wearing beads and Native American garb and by resisting all authority, especially JPL. He even tried to forcibly eject the JPL inspector who came to verify the flight readiness of his instrument. Nevertheless, he delivered his infrared spectrometers on time for both spacecraft.

During a launch simulation at Cape Canaveral in preparation for Mariner 6 a main propellant valve on the Atlas stuck open. The deflating Atlas balloon tanks sagged and began to collapse. At great personal risk, one of the crewmen ran to the pad, climbed up into the boattail, and manually closed the valve. The Atlas was damaged but the Mariner spacecraft was unharmed.[14] While that spacecraft was being checked over, the second spacecraft was mounted on a new Atlas for the first launch.

No problem in the switch, as the two spacecraft were supposed to be identical, but Pimentel expressed great alarm. It turned out that he had ignored JPL guidelines to make both of his instruments identical and had instead tried to outsmart the JPLers by tailoring his instruments to the separate Mars flyby trajectories planned for Mariners 6 and 7. Hasty changes had to be made to George's instruments in time to get both launches off within the window of opportunity.

As Mariner 6 approached Mars, Pimentel's infrared (IR) spectrometer failed to cool down as required. As the Mariner team struggled to rescue that instrument on Mariner 6 before it reached Mars, the trailing Mariner 7 suddenly went off the air, and just four days before the Mars encounter. All JPL could spare from the Mariner 6 troubleshooting team were four key engineers. Working in a tight huddle (I was watching them intently in the Space Flight Operations Facility, or

SFOF, at JPL), they studied the tracking data, somehow managed to figure out that a storage battery cell on Mariner 7 had exploded, and were able to get the spacecraft back in full operation before the flyby.

In fact, both spacecraft performed their missions well, although Pimentel only got data from his surviving instrument on Mariner 7. Based on that data alone, he declared that he had detected methane—which he concluded could only come from rotting vegetation—and therefore announced in a press conference that he had discovered life on Mars. Unfortunately, this did not stand up to challenge from his fellow scientists: the spectral lines he detected were found to be coming from solid carbon dioxide (dry ice) at the Mars south polar cap. George then modestly proclaimed himself to be the discoverer of a previously unknown spectral feature of solid CO_2.

Now, I do not mean to single out Dr. Pimentel for nonprofessional actions. He was certainly not the only successful planetary scientist who combined a strong will with a most healthy ego. Already a highly respected spectroscopist at the time of Mariners 6 and 7, his scientific career continued to be a very successful one long afterward. Later he was man enough to write Bud Schurmeier, apologizing for his behavior on the Mariner project. His story indicates some of the problems a scientific project can encounter, though. Technical problems engineers know how to tackle; it is the people problems that give them the worst fits.

The instruments and cameras on Mariners 6 and 7 revealed a very interesting atmosphere on Mars, mostly CO_2, but with some water that would migrate from pole to pole during the Martian seasons. During winter in either hemisphere water vapor would freeze first, and then, as the temperature dropped further, the much more abundant CO_2. It was winter in the Martian southern hemisphere, so the Mariners observed a large southern polar cap whose surface was composed of frozen CO_2, or dry ice. (Water ice underneath would only become visible to later missions viewing the northern polar cap during northern summer when the CO_2 cap vaporized.) The polar cap had a layered structure that was interesting to the scientists, but the surface still appeared heavily cratered, which indicated a Moon-like surface. Prospects for ever finding life on Mars appeared pretty dim at this point. It was going to take Mariner 9 to change that view.

There had now been four successful planetary missions in succession. One could certainly make the case that the golden era of planetary exploration started in 1964 with Mariner 4, but Mariners 4, 6, and 7 had the bad luck to view only ancient, cratered terrain on Mars and therefore did not portray the true dynamic history of that planet, and Mariner 5 was not equipped to return any images of what Mariner 10 later revealed to be important circulation in the clouds of Venus. The string of four successes was broken with the failed launch of Mariner 8, making the following twelve straight successes stand out even more clearly as a charmed golden era. I have therefore arbitrarily credited the Mariners of the 1960s as primarily demonstrating the technology readiness for further, more-advanced planetary missions. Thanks to the effort and inspired engineering that went into these successful projects, by 1969 launch-vehicle and spacecraft technologies were clearly ready for more ambitious planetary missions.

Determination, persistence, good engineering, and bold leadership by the pioneers of space exploration had paid off. We were now ready to stand on the shoulders of these pioneering giants and reach much higher.

CHAPTER 2

1971

MARINER 9 MARS ORBIT

By the fall of 1966 NASA had achieved success with a variety of lunar and planetary missions—Rangers, Surveyors, and Lunar Orbiters to the Moon, Mariner 2 to Venus, and Mariner 4 to Mars. Encouraged by these successes, the generous funding that came to NASA along with the Apollo Program, and the nation's determination to forge ahead of the Soviet Union in the space race, in 1966 NASA's associate administrator for Space Science and Applications, Dr. Homer Newell, and his deputy, Ed Cortright, proposed a very ambitious Mars-exploration program they called Voyager.

I entered this picture in the summer of 1967, when Oran Nicks, with whom I had worked on the Navaho 3 Project at North American Aviation and, later, the Ranger Lunar Project, talked me into taking a major cut in pay and joining him at NASA Headquarters to manage the development of the Voyager Mars Surface Laboratory. The Voyager mission was soaring in scope and estimated cost, so I did not give it much chance of surviving, but I thought that it would be very educational to work in the nation's capital for two or three years, getting firsthand experience with how the national budget process worked, and my wife agreed that it would be an interesting experience for the family. To the dismay of Cortright and Nicks some key congressmen, such as Sen. Clinton Anderson, believed that Voyager was an integral part of a multi-billion-dollar plan to land humans on Mars, and they were not about to approve that before they knew whether Apollo was going to succeed. Voyager was canceled in the fall of 1967.

A Quick Start

With Voyager down the drain, I was appointed NASA's manager of Advanced Planetary Programs and Technology. After completing Mariner '69 the Lunar and Planetary Programs Division (Headquarters Code SL) had absolutely nothing on the books, so we were starting with a perfectly clean slate. James Webb, the dynamic NASA administrator, gave us, under the heading Advanced Planetary Missions Technology (APMT), a small budget of $12 million, which we were to spread over two years, $6 million per year, just to keep JPL alive. Code SL stalwarts Paul Tarver and Milton "Mike" Mitz, two of Headquarter's finest people, helped me to organize and manage this effort. To manage its end, JPL assigned Chuck Cole, and the lab relocated the very helpful Jim Edberg (later succeeded by JPL's Norri Sirri) to Washington to assist me with future planning.

Jim Webb only reluctantly gave us $12 million, as he was unhappy with the determinedly independent manner in which Caltech and JPL director William Pickering managed JPL. Webb was also mad at Congress for failing to approve his proposed Voyager program, so he sent word down the line: "No projects, just technology development." He was going to show Congress the penalty for refusing one of his key budget requests. I was quite willing to get fired from the civil service and go back to much higher pay in industry, so I ignored his guidance and began immediately to work with JPL and the other NASA centers in designing and preparing for new planetary missions. My boss, Oran Nicks, and his successor, Don Hearth, though perhaps less willing to risk being fired, were fully supportive. We were all gambling for the good of planetary exploration.

Working with JPL, Oran and Don had already initiated planning for a 1971 mission to map the entire surface of Mars from orbit and to send probes into the Martian atmosphere as a preparatory step toward the Voyager Mars landings. Knowing that 1971 was the most favorable Mars launch year (least required launch energy) in more than a decade, we decided that our first priority after the Voyager cancellation was to restore at least the Mars orbiter mission. Of our first year's APMT budget of $6 million, I immediately allocated $1.7 million for a head start on a Mariner Mars 1971 orbiter mission and $2.7 million to prepare for our next candidate for a new start, a Mariner Venus-Mercury 1973 flyby mission.

The atmosphere was reasonably positive. Lyndon Johnson was in the White House, and he was a strong supporter of the space program. And why not? He had managed to get NASA's Apollo-leading Manned Space Center located in his home state of Texas (it was appropriately later renamed the Johnson Space Center). NASA's space science effort, including the unmanned planetary exploration program, was still being propelled by the space race with the Soviet Union and riding the coattails of Apollo—it was not yet absolutely necessary to get endorsement by the National Academy of Sciences for a new planetary mission. Homer Newell's scientific endorsement was sufficient.

Homer, who by then had been elevated to the position of associate administrator and chief scientist, not only gave his own approval but also reassured his boss Jim Webb and obtained Webb's OK to include a Mariner Mars '71 orbiting mission in the budget request being prepared for fiscal year 1969. Homer made one more contribution, which turned out to be crucial: he prudently upped our proposal to include two launches rather than one. A brilliant man, that Homer Newell. Certainly, experience dictated the wisdom of providing for two launches. Mariner 2 had recovered after the Mariner 1 failure, as had Mariner 4 after the Mariner 3 failure. Don and I had been timid in asking for only one launch, trying to sneak into the budget with a minimum-cost new start. Even with two launches, Mariner Mars '71 was a small budget item compared to the rejected Voyager Mars project, and it sailed quite easily past the White House and through Congress. We had a new start.

Organizing the Team

JPL immediately assembled a strong project team, with forceful Dan Schneiderman as project manager. Dan had been Spacecraft Systems manager on Mariner 2, the world's first successful planetary mission, and had been project manager on the successful 1967 Mariner 5 mission to Venus. By 1968 JPL had accumulated solid experience (the hard way) in coping with all the difficulties of planetary missions, and Dan was able to assemble a confident and competent team of engineers and technicians. For Mariner '71 he selected Ted Pounder as his deputy, Bob Forney as Spacecraft Systems manager (assisted by Al Conrad as Spacecraft Systems engineer), Bob Steinbacher as project scientist, Pat Rygh as Mission Operations manager, and the colorful Nick Renzetti as Tracking and Data Systems manager. Support from the Deep Space Network (DSN) was coordinated by Dick Laeser (more about his later rise to project manager in chapter 7). Another key team member responsible for mission analysis and spacecraft navigation was young Norm Haynes, who looked more like a carefree surfer from Malibu than a dedicated engineer but who was in future years to rise rapidly in responsibility at JPL. At NASA's Lewis Research Center in Cleveland, Dan Shramo was charged with shepherding the delivery of a pair of Atlas-Centaur launch vehicles from Convair.

At NASA Headquarters the demise of Voyager Mars had triggered a round of musical chairs. Homer Newell became NASA associate administrator, the number three position in the agency, John Naugle, a cosmic ray physicist, was promoted from the Division of Physics and Astronomy to replace him as associate administrator for the Office of Space Science and Applications (OSSA) with Oran Nicks as his deputy; Don Hearth became director of Lunar and Planetary Programs, and in April 1968 he brought in Dr. Donald Rea from JPL as his science deputy; and Ed Cortright was drafted to reinforce the management of the Office of Manned Space Flight (OMSF). Don Hearth immediately staffed the Mariner '71 office in Code SL with experienced Earl Glahn as program manager, Ken Wadlin as program engineer, and Hal Hipscher as program scientist.

The next step was to select the science instruments and science teams for the mission. As recommended by Don Rea and the NASA Space Science Steering Committee (SSSC) and approved by John Naugle, the following instruments, experiments, and principal investigators were chosen:

Instrument/Experiment	Principal Investigator
Television visual imaging	Harold Masursky
Ultraviolet spectroscopy	Charles Barth
Infrared spectroscopy	Rudolph Hanel
Infrared radiometry	Gary Neugebauer
S-band occultation	Arvytas Kliore
Celestial mechanics	J. Lorell

All the instruments, except for Rudy Hanel's new IR spectrometer, were flight-proven.

Design and assembly of the two flight spacecraft went smoothly. For the most part the spacecraft subsystems had either been flight-proven or had evolved from prior Mariners, although a new data-storage subsystem was developed and a large propulsion subsystem had to be added for orbit insertion at Mars. The cost tended to creep ever upward, from an early estimate of $93 million to a final grand total of $133 million, but cost growth was pretty normal for that time period, and the highly articulate and administratively adroit Don Hearth was a master at getting needed funds in the next year's budget.

Steering by the Unique James Webb

JPL needed a state-of-the-art mainframe computer to process the flood of imaging data that would stream to Earth from the two Mariner Mars orbiters. Both IBM and Univac sent proposals to build that computer, which were then evaluated by a team at JPL. Bill Pickering, JPL's director, came to NASA Headquarters to present the evaluation results to Jim Webb, who was to make the final selection. Pickering went step-by-step through the evaluation, which was totally in favor of IBM, especially in the area of software for data processing. Webb kept saying,

"Fine, IBM may be excellent, but you would say that Univac can meet your minimum requirements, right?" Pickering looked puzzled and kept repeating that IBM had the software already on line whereas Univac would have to develop it from scratch. Webb kept insisting that Univac could meet the requirements in time. Finally Bill got the message that Webb was going to give the contract to Univac and agreed that Univac could probably meet the requirements.

Several years later a young doctoral candidate at MIT by the name of Roberts wrote his thesis on biases in USAF procurements. The investigative work was cleverly done and had repercussions in the aerospace industry and in government procurement offices. On a roll, Roberts now dug into Webb's procurements at NASA, exploring anomalies and expecting to uncover a big scandal. He was surprised when Webb gave him open access to all the executive files documenting the final behind-doors deliberations leading up to contract awards. After his in-depth investigation Roberts came away a Webb fan, concluding that Jim was indeed biased in his procurements, but always for what he perceived to be the good of the nation.[1] In the Mariner Mars computer procurement Webb was concerned that IBM had a monopoly in the field and thought it best that the company have a viable competitor. He was going to help Univac become competitive.

In retrospect, Pickering proved correct. Univac did not have the software ready in 1971, and the data-processing team at JPL was hardpressed to handle imaging data from even one Mars orbiter, much less two. A small Univac computer installed in a trailer, developed for use in the field by army tactical units, had to be utilized to "predigest" the high-speed data steam so that the big Univac computer could handle at least one data stream in real time. The portable capability of the small Univac came in handy, as JPL was able to ship it to Cape Canaveral for prelaunch checkout of the spacecraft.

Drama at Launch

There were the normal number of developmental problems to be solved before the two flight spacecraft could be delivered to the cape for launch. The potting compound used in the electronic modules produced corrosion and had to be replaced with a better "glass" formulation. As anticipated, Texas Instruments struggled a bit in building Rudy Hanel's new and advanced IR spectrometer, but the instruments made it to the cape in plenty of time for integration onto the spacecraft in preparation for the May 1971 launches.

In 1970 Don Hearth had been appointed deputy director of NASA's busy and diversified Goddard Space Flight Center. I had replaced him as NASA's director of Planetary Programs[2] in June of that year. The Mariner '71 missions were the first to be launched under my responsibility, and I was really looking forward to the always-dramatic launches. Only this occasion was even more dramatic that I had anticipated. During the final countdown for the Mariner 8 launch attempt on 5 May 1971 I was seated near project manager Dan Schneiderman at the consoles in the JPL Mission Control Center at the cape (in the USAF area, actually), where I could view all the performance charts (mostly pen recorders in those days before computers and television monitors were common). As the Atlas-Centaur took off, all parameters were steady and I was relaxed; I had worked on the Atlas design, had followed its development for two decades, and had developed confidence in its reliability. My worry was about lighting up the more exotic and relatively new Centaur stage with its RL-10 hydrogen-oxygen engine.

As the Centaur separated from the expended Atlas I had my eyes glued on the recorder for Centaur-engine chamber pressure. The pen recorder showed normal ignition and leveled out at the design chamber pressure. Once fully developed liquid rocket engines are ignited they have a high probability of going full duration, so I relaxed—we were on our way. As I leaned back in my chair, Dan Schneiderman did the same. He grinned and gave me a thumbs up. With a smile I turned back to the charts. Uh oh. One of the pens had started to oscillate up and down. It was the pitch recorder. The Centaur had gone into an unstable (increasing) oscillation about its pitch axis. We watched helplessly until the Centaur finally tumbled end over end and destroyed itself. Officially, according to the State Department, it landed in the Atlantic Ocean. (Years later a charred fiberglass helium bottle that had survived reentry was found in Africa. One of ours?)

The NASA public affairs people looked around for the most senior NASA Headquarters official to face the news media people. It turned out to be me. Now I was feeling awful—almost as bad as all the folks who had put years of their lives into this project. The last thing I wanted to be was the spin doctor who would try to cast this loss in a positive light. The journalists assembled at the cape, especially some of the young female reporters, had developed real enthusiasm for this mission to Mars. As I faced the press conference I saw that these young ladies had tears in their eyes. I wanted to cry along with them. Instead, I played my management role and reminded them of all the Soviet failures in Mars mission attempts, that we had recovered with Mariner 2 after Mariner 1 failed and with Mariner 4 after a Mariner 3 failure, and that we still had four weeks in the launch window to correct whatever went wrong with this launch and get Mariner 9 properly on its way. The press conference ended on that positive note, and the Public Affairs people at NASA Headquarters were pleased with my acting ability.

Recovery

Now we had to find the source of the failure and correct it, and quickly—the sand was rapidly draining from our hourglass. We were fortunate that NASA's deputy administrator, George Low, volunteered to lead the Failure Review Team. George was the standout leader who, after the 1967 Apollo 1 pad fire that killed astronauts Virgil "Gus" Grissom, Roger Chaffee, and Edward White, headed the team that completed a thorough review of every detail of the Apollo vehicles' design and that is given major credit for the subsequent success of the Apollo Program. George was a wiry man of rather sharp features who radiated intensity. Not red-faced, shouting intensity but the cool, calm, icy type. You knew he meant business and expected results—and *now*. (I am told that George was consistently cheerful and easygoing until that fatal Apollo 1 fire, when he learned the penalty for anything less than perfection.)

When George probed for facts the contractors did not hold back. I participated as a member of the Review Team, and it was a grueling ex-

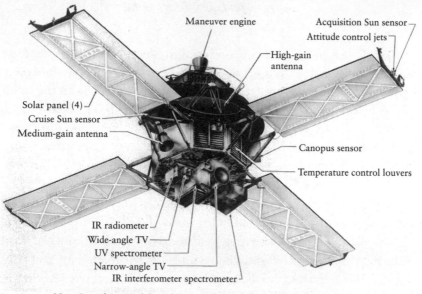

Maneuver engine

Acquisition Sun sensor
Attitude control jets

High-gain
antenna

Solar panel (4)
Cruise Sun sensor
Medium-gain antenna

Canopus sensor

Temperature control louvers

IR radiometer
Wide-angle TV
UV spectrometer
Narrow-angle TV
IR interferometer spectrometer

Note: Propulsion module and scan platform insulation blankets not shown

Mariner 9, powered by four solar panels, was launched to Mars on 30 May 1971, becoming the first orbiter of another planet. It mapped the surface, revealing Mars to have giant extinct volcanoes and a surface eroded by massive water flows some time in its wetter past. (Courtesy NASA)

perience. George worked everyone on what was for him a normal fourteen-hour day, without the slightest thought of any break over weekends, and insisted on round-the-clock laboratory testing of suspected guidance components. He ferreted out the problem (a zener diode installed backward in a guidance module), got new modules tested and delivered, and with a super effort by John Neilon's Unmanned Launch Operations (ULO) crew at the cape we were ready to launch again in just twenty-five days.

Before the failed Mariner 8 launch attempt I had treated myself to a platter of raw oysters at Captain Ed's Seafood Restaurant on the waterfront in Port Canaveral. Were the oysters some kind of jinx? I was determined not to get superstitious about launches like most aerospace people, so I intentionally went for oysters again before the Mariner 9 launch attempt. Dan Schneiderman, though, was typical when it came to jinxes: he threw away his "lucky" cap (looked like a Scottish tamo'-shanter) that he had worn during our previous failure. Maybe we neutralized our jinxes because Mariner 9 went flawlessly on its way on 30 May 1971. Nothing but smooth traces on the charts and a beautiful sight.

Mission to Moscow

From Defense Department tracking data we knew that the Russians had launched two spacecraft on trajectories to Mars, trailing our Mariner 9 by about two weeks. Through the Soviet Academy of Sciences we arranged a meeting in Moscow to explore a beneficial coordination of the three missions. The meeting began with us describing our mission plan and offering to concentrate our initial imaging in regions on Mars that were potential landing sites for their landers. Then the Russian representative got up to speak. After about fifteen minutes it became all too apparent that he was saying nothing at all—he was not answering questions or even admitting that they had launched anything toward Mars, only that they had "future plans." I became impatient and interrupted to say that we were wasting our time if they were not going to join the dialogue. Their chairman immediately called for a break until after lunch.

The senior member of our U.S. delegation was Arnold Frutkin, NASA's experienced and diplomatic head of International Affairs. He really chewed into me: "You have insulted our hosts and you will apologize immediately after lunch." I fired back, "The heck I will. They're telling us nothing." Arnold said that if I would not apologize, he would do it for me.[3] When the meeting reconvened the Russian chairman beat Arnold to the punch and immediately apologized for his delegation, saying, "Mr. Kraemer is absolutely right." He said they would now discuss their missions in detail, which they did. It turned out that before we arrived Soviet Communist Party officials had overruled their scientists and had forbidden disclosure of any information. My impulsive outburst was just what the scientists needed to break down the information barrier. (After that I was treated with great respect, and as a friend, by the Russian engineering managers and scientists. I was one of their kind.) We worked out a fine cooperative agreement to transmit Mariner 9 images of the Martian surface to them in near real time before landing attempts with their Mars 2 and Mars 3 orbiter/lander spacecraft.

Prelude to Mars

John Casani, a colorful and talented engineer at JPL, had analyzed prior troubles that we and the Russians had encountered on the way to Mars and had discussed with Donald Neff of *Time* magazine the remote possibility of a mini Asteroid Belt between Earth and Mars. Casani and Neff playfully ascribed the trouble zone to a "Great Galactic Ghoul," a name Neff immortalized in *Time*. The Ghoul must have been asleep in 1971, because our Mariner 9 and the two Russian spacecraft were healthy as they approached Mars in November.

As the news media converged on JPL for the Mars encounter, Don Hearth and Caltech scientist Bruce Murray organized a panel discussion to be held at Caltech and recorded for television. The discussion, titled "Mars and the Mind of Man," was moderated by Walter Sullivan, the highly respected science editor for the *New York Times*. Arthur C. Clarke and Ray Bradbury were there as science fiction writers and analysts of public interest in Mars, and Carl Sagan and Bruce were there to debate scientific expectations and implications. Carl and Bruce engaged in a lively debate, with Bruce taking a very conservative stand against any hard evidence for liquid water, vegetation, or any form of life on Mars. Carl had for some time proposed that the intriguing seasonal wave of darkening on Mars was due to windblown dust deposits rather than vegetation, but he held out considerable hope for moisture and the evolution of micro- or macro-organisms. Toward the conclusion of the panel discussion Ray Bradbury said that he had written a poem he thought appropriate for the occasion. It was about human yearning for higher things, for greater knowledge and even immortality, and as he neared the end of his poem, Ray spoke with increasing animation:

IF ONLY WE HAD TALLER BEEN
The fence we walked between the years
Did balance us serene;
It was a place half in the sky where
In the green of leaf and promising of peach
We'd reach our hands to touch and almost touch that lie,
That blue that was not really blue.
If we could reach and touch, we said,
'Twould teach us, somehow, never to be dead.

We ached, we almost touched that stuff;
Our reach was never quite enough.
So, Thomas, we are doomed to die.
O, Tom, as I have often said,
How sad we're both so short in bed.
If only we had taller been,
And touched God's cuff, His hem,
We would not have to sleep away and go with them
Who've gone before,
A billion give or take a million boys or more
Who, short as we, stood tall as they could stand
And hoped by stretching thus to keep their land,
Their home, their hearth, their flesh and soul.
But they, like us, were standing in a hole.

O, Thomas, will a Race one day stand really tall
Across the Void, across the Universe and all?
And, measured out with rocket fire,
At last put Adam's finger forth
As on the Sistine Ceiling,
And God's great hand come down the other way
To measure Man and find him Good,
And Gift him with Forever's Day?
I work for that.
Short man. Large dream. I send my rockets forth between my ears,
Hoping an inch of Will is worth a pound of years.
Aching to hear a voice cry back along the universal Mall:
We've reached Alpha Centauri!
We're tall, O God, we're *tall!*

As he recited the last line an emotional Ray leaped to his feet and raised both arms to the heavens. Normally blasé Caltech professors and grad students rose in a spontaneous standing ovation that lasted for many minutes. It was a magic moment. Ever since then Ray has been adopted by planetary scientists as the poet laureate of planetary exploration. Bruce Murray was not alone when he wrote that the poem expressed how he felt and still feels. In Bruce's words, "Long after our burned-out bodies have turned to dust, when our individual names are remembered only by genealogically minded descendants, our first look at our planetary neighbors will be collectively remembered. Faintly, we too will survive as part of humanity's future—and outwit, a little, our own mortality."[4]

The Slow Unveiling

In September 1971 a yellow cloud was observed through telescopes to be forming in Mars' southern hemisphere. Within two weeks this dust cloud had spread over the entire planet. Mars' orbit around the Sun is somewhat elliptical, and 1971 marked its closest approach to the Sun. The increased solar intensity generated winds estimated at more than 645 kilometers (400 miles) per hour and stirred up the greatest Martian dust storm ever observed. On 14 November Mariner 9 made it safely into Mars orbit (the first spacecraft to orbit another planet) only to be disappointed by views of just dust clouds. Fortunately, all subsystems were strong and we could afford to wait for the dust to subside. The Russians were not so fortunate: their two orbiters were pre-programmed and expended all their image data storage before the dust cleared, and their two landers both perished trying to land in that wind. Our barrier-breaking U.S.-USSR cooperative agreement was all for naught.

In the first Mars images returned from Mariner 9 the only features distinguishable through the clouds were the white south polar cap and four puzzling dark spots in the dust clouds north of the equator. As the storm gradually subsided and the dust settled it became apparent that the dark spots were the tops of enormous volcanoes. The largest, already named Olympus Mons by astronomers, towered more than 21.7 kilometers (70,000 feet) above its surroundings (or almost 90,000 feet

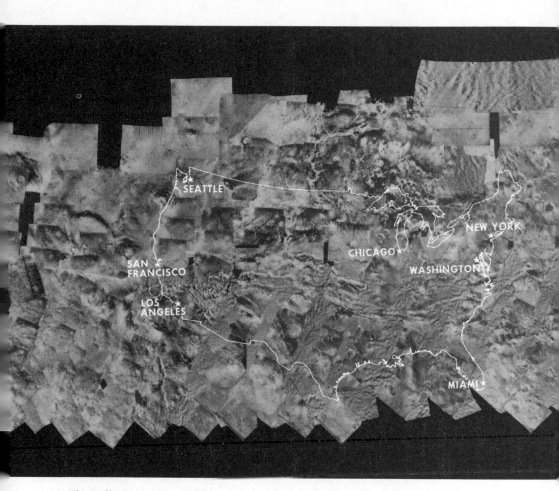

The Valles Marineris on Mars as viewed by Mariner 9. This "Grand Canyon" of Mars is seven kilometers (four miles) deep, and with a same-scale map of the United States superimposed on it, it stretches from New York to San Francisco. (Courtesy NASA)

above the mean Martian surface), more than three times the height of the tallest peak on Earth (not Mount Everest, but rather Mauna Kea on the big island of Hawaii as measured from its base on the ocean floor). But the best was yet to come. Streaming from the volcanoes were immense lava fields that had covered earlier cratered terrain. The incoming images were organized in mosaics, revealing down by the equator east of Olympus Mons a spectacular "Grand Canyon of Mars," long enough, at almost 4,800 kilometers (3,000 miles), to cross the entire continent of North America from New York to San Francisco and a yawning 6.5 kilometers (4 miles) deep. The steep walls of this canyon were marked by immense landslides. This striking canyon was quickly and officially named Valles Marineris in honor of its Mariner discoverer.

The Sagan-versus-Murray debate resumed with conservative Bruce trying to explain the canyon and its surrounds as the result of the eroding action of strong winds. Yet as more and more images were added to the mosaic, showing tributaries feeding into the canyon and then braided channels downstream sweeping northward in a great curl to empty into a vast alluvial plain, the direct comparisons to Earth left little doubt that these features were caused by massive water flow. Bruce gracefully backed down.

With all the water on Mars existing now as either vapor in the atmosphere or solid ice, where did all of that liquid water come from? The leading explanations were (1) that Mars had more water in the past or, more likely (2) that the active volcanic period melted all of the water locked up as permafrost in the soil and produced a geologically brief period of massive flooding. Life-forms could have evolved then and might still be surviving at today's permafrost boundary. The news people were enthralled with Carl's speculations. In his words, "Some ancient channels seem to have been carved by rainfall, some by underground sapping and collapse, and some by great floods that gushed up out of the ground. Rivers were pouring into and filling great thousand-kilometer-diameter impact basins that today are dry as dust. Waterfalls dwarfing any on Earth today cascaded into the lakes of ancient Mars. Vast oceans, hundreds of meters, perhaps even a kilometer, deep may have gently lapped shorelines barely discernible today."[5] The man had a lovely way with words.

Hal Masursky was a star at the press conferences in his vivid and knowledgeable interpretations of the geological features on Mars. In addition to being a leader in his field, Hal also had a talent for clear verbal descriptions. His enthusiasm was infectious. In fact, we had to stay alert to that enthusiasm—Hal was diabetic and would get so involved with the new images coming in that he would ignore his diet and might either forget his insulin shot or mistakenly do one twice. If he suddenly ran out of steam at a meeting it did not mean that he was just tired. He was probably going into insulin shock and needed a quick shot of orange juice. Hal wound up with a number of watchful "mother hens" among his many friends on the Mariner Science Team.

Other panelists included the youthful Noel Hinners, lead geologist and Headquarters scientist for the Apollo Program, who gave interesting presentations in which he compared the craters on Mars with those on the Moon. Conway Leovy explained the formation and disappearance of the dust clouds. As they cleared it was apparent that the dust layer had brightened the surface. Then, as surface winds gradually redistributed the dust, the dark areas familiar to astronomers reappeared, vindicating Carl Sagan's earlier prediction. New findings by PIs on the composition and properties of the Martian atmosphere and temperatures over the surface were interesting and scientifically important. Humidity and vertical profiles of temperature were monitored at different latitudes and seasons. The polar ice caps were confirmed to be predominantly frozen carbon dioxide. Despite the importance of these findings, none of them could compete with imaging when it came to public interest.

I was personally pleased with all the comparisons of Mars to Earth and the Moon being made—what has come to be known as "comparative planetology." As I noted in my 16 February 1971 testimony before Rep. Joseph Karth's House Subcommittee on Space Science and Applications,

> During the past decade we have been able to explore only Venus and Mars. These limited efforts have already yielded a new understanding of the origin and evolution of Earth and its neighboring planets. One current theory holds that Mercury, Venus, Earth, and Mars had similar beginnings but were influenced by differing environments so that they are currently at different stages of evolution. Mercury's proximity to the Sun speeded its outgassing and loss of its atmosphere to space. Venus evolved slower, but enough time has passed for Venus to have had substantial outgassing from its surface, with essentially all the lighter gases having escaped into space. The result was a progressive buildup in the carbon dioxide content of the atmosphere. Carbon dioxide (CO_2) tends to trap heat from the Sun, even-

tually producing a "runaway greenhouse" effect where trapped heat liberates more CO_2 from surface materials, which in turn traps even more heat. As a result, Venus has now evolved to the state of a very heavy atmosphere (90–100 times Earth's) and searing surface temperature (800–900 degrees Fahrenheit).

On Earth, in contrast to Venus, a favorable set of conditions allowed the early evolution of plant life. These living organisms provide the unique function of converting CO_2 back into oxygen. About Mars we are much less certain. It may or may not be evolving the life-forms that could influence its future evolution.

By collecting further data on these neighboring planets we hope not only to improve our understanding of the origin and evolution of the solar system, but also to improve our ability to predict how Earth might react to changing conditions. For example, the further study of Mars and Venus should increase our understanding of whether a continued dumping of exhaust products and reduction of vegetation could eventually lead to a "runaway greenhouse" effect on Earth that could endanger all future life on our planet. Certainly we must give highest priority to further study of Earth itself for these vital answers, but many scientists believe that our neighboring planets can supply important inputs to the understanding of Earth's dynamic phenomena, such as weather, climate, earthquakes, and changing magnetic field.

At NASA Headquarters geologist Steve Dwornik expressed my feelings well when he wrote, "The exciting thing about comparative planetology is that it will permit us to unfold the lost part of the Earth's history, now largely obliterated by erosion, mountain building, and other processes. . . . Comparative planetology is the starting point for an understanding of the physical future of planet Earth."[6]

Mars Comes Alive

The Mariner 9 mission marked a dramatic escalation of interest in Mars. After the 1965 Mars flyby of Mariner 4 showed ancient cratered terrain, Oran Nicks had lamented, "The dashed hopes of finding 'little green men' was devastating to the support for Mars exploration—especially from administrators and members of Congress."[7] The subsequent Mariners 6 and 7 only reaffirmed a Moon-like, dead-looking surface. It had been pure bad luck that Mariner 4 imaged a narrow swath from north to south that included only cratered areas. The imaging on Mariner 6 and 7 was constrained by the flyby trajectory to include only the southern hemisphere, which is mostly cratered. The equatorial region and the northern hemisphere were more volcanic, however, and the northern hemisphere, being generally lower than the southern hemisphere, is where the greatest water flooding was concentrated. When Mariner 9 mapped the entire planet at one-kilometer resolution, and selected areas at one-hundred-meter resolution, Mars came alive. Author Clayton Koppes summarized it nicely: "The northern hemisphere fairly blossomed with evidence of recent geological and meteorological activity, some of which was analogous to that on Earth."[8]

Any lagging in the desire to land on Mars was reversed practically overnight. We were ready to push on with the ambitious Viking Mars Landing Project (coming in chapter 6).

CHAPTER 3

1972 AND 1973

PIONEER 10 AND 11
JUPITER AND SATURN FLYBYS

As we continued our ride on the coattails of Apollo and the space race, getting the first two Pioneer planetary missions into the NASA budget was relatively easy, although it did take a good deal of rather devious—you might even say sneaky—administrative maneuvering. But first, a little Pioneer history and some necessary wading through the maze of spacecraft alphabetical and numerical designations.

In 1958 and 1959 four small Pioneer spacecraft were built by the USAF Space Technology Laboratories and JPL and launched with limited success toward the Moon. A fifth Pioneer, built by NASA's Goddard Space Flight Center, was launched in March 1960 and successfully went into orbit around the Sun. Then early in the 1960s NASA's Space Physics and Astronomy Programs Division (Headquarters Code SP) was granted approval for a series of seven Pioneer interplanetary missions to study the solar wind. For the first missions of the series, five small, spin-stabilized cylindrical spacecraft were built by TRW under the management of NASA's Ames Research Center (with Charles P. Hall as project manager) and launched from 1965 to 1968. Following the NASA custom of labeling spacecraft alphabetically before launch and then numerically after launch, the new Pioneers A, B, C, and D became

Pioneers 6, 7, 8, and 9 after launch. Pioneers 6 to 9 were spaced around the Sun at Earth's orbital distance from the Sun, where they set all-time records for long life, functioning for more than two decades, and provided the Apollo astronauts with advance warning of dangerous solar flares. Pioneer E, fifth of the Ames-TRW series, was lost in the destruct of its Thor-Delta launch vehicle and was not given a numerical designation.[1] The last two of the interplanetary Pioneers, F and G, were to be modified designs to track the solar wind to 4 AU (one AU, or astronomical unit, equals the mean distance between Earth and the Sun).

In 1967, when the ambitious Voyager Mars program was canceled, Code SL's only mission was to complete the 1969 Mariner Mars flybys. To give the capable people in Code SL something to manage the Pioneer F and G missions were transferred to them from Code SP. It did not take long for the new director of SL, Don Hearth, and his manager of Advanced Planetary Programs, your author, to decide to heck with 4 AU, let's go to 5 AU and fly by Jupiter. This sounded like a simple targeting change, so overnight we had a new planetary exploration project without having to go through the usual series of approvals.

We already had good scientific support for missions to Jupiter. By 1967 Jim Van Allen was pushing hard within the NASA advisory Lunar and Planetary Missions Board (LPMB) for low-cost missions to the outer planets. It took very little prodding to get a formal endorsement from the Space Science Board (SSB) of the National Academy of Sciences, which in June 1968 recommended that "Jupiter missions be given high priority, and that two exploratory probes in the Pioneer class be launched in 1972 or 1973." Don Hearth had all the backing he needed, and in February 1969 NASA Headquarters officially approved for Code SL the Pioneer F and G missions to Jupiter.

Various groups leaped at the opportunity to build the first spacecraft to explore Jupiter—clearly the dominant planet in the solar system, with a mass twice that of all the other planets combined and a family of large satellites that constituted a miniature solar system. Scientists and engineers at NASA's Goddard Space Flight Center had always wanted to manage a planetary mission, and under the experienced leadership of Rudy Stampfl and Emil Hymowitz they came to SL with quite detailed studies of a spacecraft concept they called the Jupiter Galactic Probe. It featured a large antenna dish and was powered by radioisotope thermoelectric generators (RTGs). Their artist's painting in full color, which they gave to me and I still have, is almost an exact image of the final Pioneer 10/11 design.

Meanwhile, at Ames, the quiet-spoken but highly capable Howard Matthews was busy detailing the Jupiter mission opportunities and requirements. In addition, TRW geared up an impressive study effort headed by their new-missions guru, Dr. Herb Lassen. Tall and with a trim beard, Herb was an impressive figure. He was technically very sharp and creative—and quite a marketeer. He came to SL with a detailed briefing showing that a fine Jupiter mission could be performed with a straightforward upgrade of the reliable Pioneer 6 to 9 spacecraft design and for the bargain price of only $20 million. Having been associated with similar sales pitches as a contractor, I immediately doubled that figure in my mind to $40 million.

Even though Herb Lassen thought the Jupiter spacecraft could be powered with extended solar panels, I figured that we would probably wind up with something like GSFC's RTG-powered Jupiter Galactic Probe. It looked like a new spacecraft design to me, meaning that we would have to go through an industrywide competition. Don Hearth had other ideas, however. Don, a true master of working within government bureaucracy, knew that at Ames he had a strong project manager in Charlie Hall, who worked well with the talented engineers at TRW. Somehow Don convinced his bosses at NASA and all the by-the-book NASA procurement people that Pioneers F and G were a straightforward evolution of the Pioneer A to E series and that the two new spacecraft should just be negotiated into Ames' ongoing Pioneer contract with TRW. "Sole source," no competition. This arrangement negated any management role for GSFC.

I know I would not and could not have pulled off this maneuver, but Don did, and as a result he advanced these missions by a good two years or more. In truth, he also probably saved both the government and industry a good deal of wasted effort and dollars as Don would have given NASA management responsibility to Ames and it is likely that TRW would have won any competition conducted by the Ames team for the new Pioneer spacecraft.

Assembling the Pioneer Team

Management practice at TRW was to keep their sharpest engineers on proposal-writing teams in order to bring in new contracts. William J. "Bill" Dixon, a thin, bespectacled systems engineer, was the "genius" working under Herb Lassen to define the total systems features in the new spin-stabilized F/G spacecraft. After contract negotiations were started TRW wanted to pull Bill to go to work on other proposals, but I believed that only Bill knew all the design interactions, such as how to balance and dampen the spacecraft so it would spin without wobble or precession. I insisted that Bill stay on the project as Spacecraft Systems manager. It may have been micromanagement, and it certainly was a strong-arm move, but I never regretted doing it. Bill and TRW's likable project manager, Bernard J. "O. B." O'Brien, made an excellent team and worked smoothly with their counterparts at Ames.

Without debate the key man at Ames was Charlie Hall. Charlie was about as close as you can get to my image of an ideal project manager for a project like Pioneer. His straightforward, unpretentious manner generated instant confidence and liking for the man. He obtained the facts, made timely decisions and assignments, and without raising his voice or pounding the table made it quietly clear when he expected actions to be completed. His boss, John V. Foster, and the center director, Hans Mark, wisely did not try to micromanage Charlie, even though at times he appeared to be doing too much on his own. For example, he was in fact if not in title his own Spacecraft Systems manager. He handled it easily, though, and drew dedicated and loyal support from all members of his small project team.

Initially Charlie's day-to-day Headquarters interface within SL was NASA veteran Glenn Reiff, who was also program manager for NASA's share of the joint U.S.-German Helios Project. When Glenn decided to move to Colorado and transfer to the Department of Transportation I had to find a quick replacement. At Ford Aeronutronic I had gotten to know Fred Kochendorfer, who had served as NASA's program manager for early Mariner missions before going to work for Ford managing communications satellite projects. I was able to talk the experienced Fred into returning to NASA Headquarters to replace

Glenn Reiff. As I expected, Fred worked very well with Charlie Hall. Al Opp, a top-notch physicist from Code SP and a great person to work with, served as our program scientist.

NASA's Space Science Steering Committee, aided by Hearth's highly respected science deputy Don Rea, recommended fourteen experiments for the missions. John Naugle, as associate administrator for Space Science and Applications, approved all fourteen, along with their principal investigators:

Instrument/Experiment	Principal Investigator
Magnetic fields	Edward J. Smith
Fluxgate magnetometer	Mario H. Acuna
Plasma analyzer	John H. Wolfe
Charged particle composition	John A. Simpson
Cosmic ray energy spectra	Frank B. McDonald
Jovian charged particles	James A. Van Allen
Jovian trapped radiation	R. Walker Fillius
Asteroid-meteoroid astronomy	Robert K. Soberman
Meteoroid detection	William W. Kinard
Celestial mechanics	John D. Anderson
Ultraviolet photometry	Darrel L. Judge
Imaging photopolarimetry	Tom Gehrels
Jovian infrared thermal structure	Guido Munch
S-band occultation	Arvydas J. Kliore

Although GSFC did not win the Pioneer F/G project management role, a number of Goddard scientists were selected as members of the science teams, including PIs Mario Acuna and Frank McDonald.

Project Escalation

TRW quickly gave up on the use of solar panels. Jupiter is five times farther from the Sun than is Earth, and solar intensity diminishes as the square of distance from the Sun, so at Jupiter a spacecraft's solar panels receive only one-twenty-fifth the energy at Earth's orbit. At this low intensity it was found that available solar cells might not generate any useful electricity at all. The switch to RTG power meant a change in design, and the cost went up.

The cost would have escalated even more but for the fact that the Atomic Energy Commission provided us with the RTGs at no cost. In 1967, when I moved from California to the Washington, D.C., suburbs, I found that my across-the-street neighbor was Glenn Newby, who directed the AEC programs in radioisotope devices. I invited Glenn to come spend a Saturday with us at NASA Headquarters and we dazzled him with color slides of the outer planets and promising future exploration opportunities, all of which would only be possible with RTG power. Glenn was swept up in our enthusiasm and worked with NASA's Harry Finger to convince AEC top management (over the vigorous protest of their comptroller) that it was in the AEC's best interest to not only develop RTGs for spacecraft but also to provide the flight units at no cost to NASA. This favorable cooperative agreement held up for the later RTG-powered Viking Lander and Voyager missions.

During the 1960s cost overruns on NASA projects were pretty much accepted. Within OSSA the Surveyor Project had required a major bailout. By comparison the Mariner projects had done much better, and I had watched Don Hearth procure funds quite readily for a modest overrun on Mariner Mars 1971. Although Pioneer F/G had grown from $40 million to almost $100 million, that total included years of operations and data analysis so it still looked like a bargain and was not seriously threatened. However, when I was appointed director of SL in 1970 I was determined to put a dollar cap on all our flight projects. One of the reasons, I suppose, was purely pride—while employed in the aerospace industry I had a solid, unblemished (and a bit lucky) record of completing difficult research and development (R&D) projects within budget. I wanted to continue that reputation. More important, I could see the NASA budget getting rapidly tighter as the Apollo missions wound down. Future overruns would have to be covered by cutting back or even canceling other projects. If a large planetary project were to cause the cancellation of a smaller science mission, the space science community would be down on new planetary programs for many years to come. We had great plans for many future planetary missions, but project cost overruns could doom it all. So I was determined to stop further cost growth on the Pioneer Project.

As the F and G spacecraft were being assembled at the TRW plant in Redondo Beach, California, O. B. O'Brien and his crew proposed a rather startling manpower plan. It fit with TRW's overall manpower availability for them to perform an intensive effort to complete assembly and test of the proof-test spacecraft and the two flight spacecraft earlier than originally planned and then store them prior to launch preparations. They argued that this would save money by letting them get people off the project earlier. The problem was that it resulted in a planned manpower curve that peaked at a high level and then dropped precipitously. The headcount-versus-time curve looked like the profile of the Matterhorn. If the peak level of manpower persisted for even two or three weeks it would blow our budget to smithereens. I persuaded the normally mild-mannered and soft-spoken John Naugle to join me as we marched into TRW headquarters and addressed their top management, threatening them in no uncertain terms with never getting another NASA contract if they blew that manpower plan. Strong-arm tactics, but maybe it helped because TRW followed their manpower plan perfectly and completed the flight spacecraft right on time and for a bargain price of $38 million.

Dan Goldin, who was to become NASA administrator in the 1990s, was a member of TRW's design team on Pioneer. When I introduced myself to him years later when he reported as administrator, he said, "Oh, I remember you. You were that SOB from NASA Headquarters that pounded us bloody on Pioneer costs." That is a poor way to be remembered, but he was correct.

A Bumpy Path to a Launch

There were the usual crises during spacecraft fabrication and testing. During one of the final environmental tests of the first flight spacecraft, for example, a lifting cable somehow came unhooked, causing the spacecraft to swing down and hit the sharp lip of TRW's large thermo-vacuum tank. The impact broke a major structural member, one of the struts attaching the large nine-foot-diameter antenna to the body of the spacecraft. In order to save weight the strut had been fabricated from exotic boron filaments embedded in epoxy rather than from metal or conventional fiberglass. There were no spare struts, so a repair was attempted, although there were concerns about how a repair with more epoxy would work. To everyone's great relief (and my surprise) the repaired strut tested out stronger than before the accident.

The Pioneer team at Ames was having troubles of its own in getting all the science instruments delivered to TRW on schedule. Joseph Lepetich, working for Charlie Hall, did a superb job prodding those instruments along in university labs and at subcontractors all over the country. Some of the PIs grumbled to NASA Headquarters about Joe's quiet but firm pushing; but Joe was arranging for added help for the PIs as well as just pushing, and I give him great credit for getting all the instruments delivered on schedule. Ralph Holtzclaw was in a similar role for Hall, working with TRW on the spacecraft hardware, and Bob Hostetler was teamed with Dan Shramo at NASA's Lewis Research Center to keep our Atlas-Centaur launch vehicle on schedule.

James Schlesinger, later appointed secretary of defense, was at that time heading the AEC, which so generously supplied our RTGs. He had expressed great interest in our Pioneer Jupiter missions, so NASA's administrator, Jim Fletcher, invited him down to Cape Canaveral to witness the launch of Pioneer F in late February 1972. Because we were heading outward in the solar system on a direct trajectory without a parking orbit, the launch would be in the evening so that in progressing eastward the spacecraft would leave Earth in the same direction as Earth's rotational velocity around the Sun. Thus the velocity of Earth's rotation about its axis would add to the velocity of Earth's movement around the Sun. These night launches are a beautiful sight, even when viewed from miles away. Illuminated with powerful floodlights, the shining silver and white launch vehicle stands out starkly, with almost unearthly brilliance, against a black sky over the Atlantic Ocean. Even before the added fiery tail of the rocket exhaust, it is a memorable sight. We hoped to give Schlesinger and Jim Fletcher and the hundreds of spectators a visual treat to remember. Not to be.

The countdown to launch was going smoothly, but balloon-borne radiosonde measurements of upper-atmosphere winds were indicating an increasingly severe wind-shear layer. The Atlas and its tissue-paper-thin balloon tanks are amazingly rigid when pressurized, but they have their limit. As the countdown progressed the last radiosonde was released and indicated a shear magnitude right at our safe structural limit. The launch vehicle people from NASA Lewis were accompanied by their director, Bruce Lundin. Charlie Hall and I turned to Bruce, expecting him to say, "No go." Instead he started to waffle and confer with his people as to whether we "might" proceed. I did not like this indecision one little bit and was relieved when Charlie ruled that we were going to stand down for the night. We then descended on the Lewis team and pressured them to set firm criteria on when it was safe to launch and then stick to them.

That scrub was on a Sunday. Monday was the same—high wind shear. On Tuesday the launch range was reserved for another launch. Wednesday, more high winds. Thursday, 2 March 1972, Pioneer F (now relabeled Pioneer 10) rose from the pad at 8:49 P.M. eastern standard time. Due to the delays we had lost many of the VIPs among our spectators, such as Schlesinger and Fletcher, but our bird was in flight. Sent on its long journey atop a one-and-a-half-stage Atlas, a Centaur high-energy upper stage, and a Thiokol TE-364 solid-propellant kick stage, Pioneer 10 traveled at a record-breaking speed of 51,713 kilometers (32,114 miles) per hour, almost 7,000 miles per hour faster than achieved by any previous manmade object and fast enough to pass the Moon's orbit in just eleven hours rather than the usual three days.

A Perilous Passage

After separation from the upper stages the spacecraft's RTGs were released to deploy to the ends of their two long booms. There was brief concern when one of the RTGs failed to confirm that it had locked into place, but that was soon confirmed. Then the even longer 5.2-meter (17-foot) magnetometer boom was extended, and the spacecraft settled into its long journey.

The next big hurdle was the Asteroid Belt, a band of miniplanet bod-

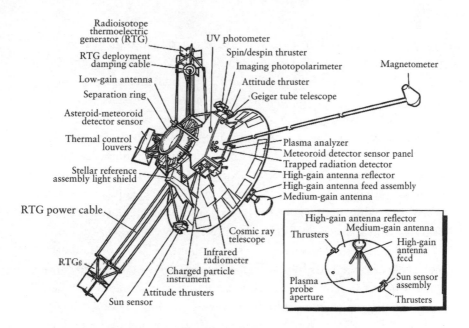

The Pioneer 10/11 spacecraft, powered by RTGs and spin-stabilized to keep
its large, high-gain antenna pointed at Earth. Pioneer 10 was launched on 2
March 1972, making the first flyby of Jupiter and becoming the first manmade object to head out of the solar system. Its twin, Pioneer 11, was
launched on 5 April 1973, viewed the south pole of Jupiter, then made the
first flyby of Saturn. (Courtesy NASA)

ies in orbit around the Sun between the orbits of Mars and Jupiter.
There are thousands of these minor planets or asteroids in the Belt, and
more are regularly being discovered by astronomers. Ceres, the largest
asteroid observed to date, is large enough in diameter to blanket the
state of Texas. Pioneer 10 could be steered to avoid the observed asteroids, but what about smaller, unobserved rocks and debris? It was
likely that repeated collisions among the asteroids had generated great
quantities of sandlike particles that could not be detected with even the
best telescopes, and at the tremendous speed of encounter, even a grain
of sand could put a substantial hole in our precious spacecraft. So there
was no way of knowing whether a spacecraft could survive a passage
through the Asteroid Belt. Pioneer 10 would find out.

One of the PIs, Bill Kinard, had mounted 234 small pressurized pil-
lows on the back of the antenna. When a pillow or cell was punctured

by a particle in space it lost gas at a rate proportional to the size of the hole made in the cell wall. Pioneer 10 not only made it through the Asteroid Belt unharmed, but Kinard found no unusual increase in particle density in the Belt region even though he measured impacts with particles as small as one-billionth of a gram. This removed a major concern about a potential barrier to the exploration of the outer solar system.

Charlie Hall had Pioneer mission operations going smoothly at Ames under his managers Robert "Skip" Nunamaker and Richard Fimmel. Fimmel especially was a pleasure to watch at work. He always seemed to have everything under control and to be thoroughly enjoying the adventure of his spacecraft. To better appreciate his calm demeanor, one must realize that the Pioneers are so simple that they need constant care like a baby. Unlike JPL's Mariners, the Pioneers have no onboard computer to direct functions in the absence of uplink commands or when things go wrong; they must be told every little thing to do by direct command from Earth. It took sixteen thousand commands to get Pioneer 10 past Jupiter. One bad command and you might kiss the spacecraft goodbye forever. You had to remember to keep repointing the spin axis, too: for the best data return the antenna needed to be pointed within .5 degrees of Earth, which was constantly moving in its path around the Sun. Fimmel, Nunamaker, Norm Martin, and the rest of the ops team had to constantly prepare and check new commands to be sent to the tiny craft at all hours of the day and night. And the busy but always upbeat Richard Fimmel smiled through it all.

First to Jupiter

As Pioneer 10 approached Jupiter the Ames Research Center became a real carnival. It was Ames' first experience with a planetary encounter. Their cool public affairs officer, Pete Waller, had things well in hand, but I think most of the center personnel were astounded at the number of reporters and television crews that descended on their relatively small facility. It was really great for morale, even for those just tightening bolts in the Ames wind tunnels and not connected in any direct way with Pioneer.

Pioneer 10 had no camera aboard but could reconstruct images from Tom Gehrels's imaging photopolarimeter. As the spacecraft spun, the instrument could scan contiguous strips .03 degrees wide from which images could then be assembled. The instrument recorded in red and blue wavelengths, to which a judicious amount of green had to be added to give the true color of Jupiter. The color images clearly showed the turbulent bands in Jupiter's clouds and the dramatic Giant Red Spot cyclone in detail never before observed. The particles and fields instruments began to paint a picture of an enormous rotating magnetic field buffeted by the solar wind and with complex interactions with the Galilean satellites (the four large moons of Jupiter discovered by Galileo with his first telescope). The news people were fascinated. Every morning Charlie Hall and his project scientist, John Wolfe, enthralled the press conference audience with a short but clear summary of the spacecraft status, the overnight results, and the plans for the next twenty-four hours. These briefings became known affectionately by the news media as the "John and Charlie Show."

Material for the briefings came from stand-up meetings in Charlie's office first thing every morning. This management tool of Charlie's was absolutely great. Every PI and manager gave a brief but concise report; having everyone standing, crammed into a small office, motivated all concerned to be brief. All hands were brought up to date and, most important, each PI heard the results from all the other experiments, which greatly aided the interpretation of their own data. Charlie said that he got confirmation of the value of those meetings when one morning he arrived at the office especially early to find the two top scientists from NASA Headquarters, John Naugle and Homer Newell, camped outside his office door to make sure they would be able to squeeze into the stand-up session.

Challenging Mighty Jupiter

On 3 December 1973, as Pioneer 10 neared its closest approach to Jupiter, the radiation-intensity readings began to rise steeply. The Pioneer team had targeted their spacecraft to pass only 1.86 Jupiter radiuses above the Jovian cloud tops, and now it looked like they had been too daring. The science experiment teams were getting data dumps every half hour, and with each new reading of rising radiation levels the mood became more gloomy. Charlie called a special stand-up meeting in his office. The normally bright-eyed and bouncy Van Allen gloomily said that he did not expect another useful data dump—his charged particles instrument would be saturated. The intensity of high-energy electrons was one thousand times what had been predicted. The TRW engineers said flatly that Pioneer 10 was about to go off the air. Everything on the spacecraft was being fried.

I walked back with Van Allen to his team cubicle to await the next hoped-for data dump. The mood was grim indeed. When the data sheets came in, surprise! Not only were there valid data but the radiation readings were *down!* And they continued to go down. As soon as the magnetometer data were analyzed the reason became clear. Jupiter's dipole magnetic field was found to be tilted eleven degrees from the axis of rotation. That means that as the planet rotates about its axis the radiation belts wobble up and down, like a flat tire that is skewed in its mounting on a wheel. By a great stroke of good luck the radiation belt swung up out of the way just in time to let Pioneer 10 go streaking through unharmed. Well, mostly unharmed, because the tiny craft had still absorbed a dose of radiation ten thousand times the level that would be lethal to humans.

At the press conference immediately after encounter I wanted to dramatize just how lucky we had been that our spacecraft had survived. What came to mind was a test a young physicist (I believe his name was Shapiro) used to do at Los Alamos during the development of the first atomic bomb. The bomb's trigger mechanism required two small radioactive hemispheres to come together to start the chain reaction. To test their potency Shapiro used to set the two hemispheres well apart on a work bench in front of a Geiger counter and then start to nudge the hemispheres together with a screwdriver. When the Geiger counter started to tick like crazy he would deftly flick the hemispheres apart before he received a dangerous dose of radiation. Risky stuff. The procedure became known as "tickling the dragon's tail." So at the press conference I said, "We can say that we sent Pioneer 10 off to tweak a dragon's tail, and it did that and more. It gave it a really good yank and it managed to survive." Carl Sagan would have expressed it much better, I am sure, but the press did pick up the phrase and it made it into books written later about the mission.

With all the media attention the project was getting there was a lot of coverage given to the gold-anodized aluminum plaque mounted on both spacecraft as a message for extraterrestrials who might recover the craft—sort of a message in a bottle set adrift on the vast sea of space. Carl Sagan, his artist spouse Linda Salzman Sagan, and Frank Drake had designed this plaque to give any intelligent species some decipherable information on when the spacecraft had been launched, where it had originated, and what humans looked like. At the bottom of the plaque the point of origin, Earth, was indicated as the third planet out from our star, the Sun. The location of the Sun with reference to other stars in our galaxy was indicated by rays pointing to known pulsar radio sources, with the binary numbers at the ends of the rays representing the pulsars' frequencies at the time of launch relative to that of the hydrogen atom shown at the top of the plaque. Because pulsar frequencies decrease over time, an extraterrestrial investigator could deduce when the spacecraft was launched. A side profile of the Pioneer spacecraft (added at the suggestion of Skip Nunamaker) gave a yardstick for the height of two human figures, an unclothed man and a woman.

When the plaque design was submitted to NASA Headquarters for approval I must confess that I was a bit nervous about it. Linda was a skilled artist and her naked human figures were very detailed and realistic, as they needed to be. It seems a bit silly today, but at the time I feared that some taxpayers, the true owners of the spacecraft, might label it pornographic. My boss, John Naugle, had no such fears and approved the design but with the one compromise of erasing the short line indicating the woman's vulva. (The poor extraterrestrials are probably going to be puzzled by the functional differences in anatomy between the two figures.)

We received a few crank letters, but my fears of widespread public outcry were unfounded. The *Los Angeles Times* published a cartoon showing a Pioneer spacecraft crashed on a cratered planet and a nicely

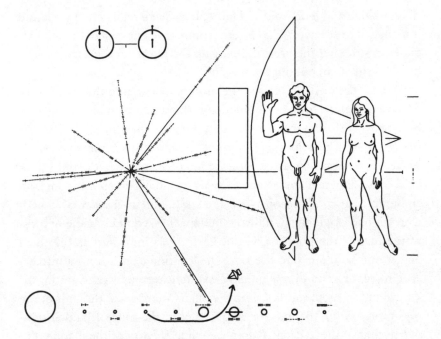

Pictorial image etched onto gold-anodized aluminum plaques on the Pioneer 10 and 11 spacecraft to indicate to the inhabitants of other star systems the location and nature of the spacecrafts' human creators. Launch time and location of our solar system are indicated by rays pointing to known pulsar radio sources, the binary numbers at the ends of the rays representing the pulsars' frequencies at launch time relative to that of the hydrogen atom, shown at the top of the plaque. (Pulsar frequencies decrease over time, enabling an extraterrestrial to deduce when the spacecraft was launched.) (Courtesy NASA)

dressed couple standing near it examining the plaque, the man saying, "It appears that Earth creatures are much like us, except they don't wear any clothes." After seeing that cartoon I relaxed, with increased faith in the common sense of the American public. I learned later that there had been one more alteration of Linda's art work. The public affairs people at Ames, thinking the facial features on her figures too ethnic, neutralized them into a racial mixture. I guess I never saw Linda's original art work after all.

On to Saturn

Pioneer 10's twin, Pioneer G, was launched at 9:11 P.M. eastern standard time on 5 April 1973 and was relabeled Pioneer 11. Because it was trailing Pioneer 10 by thirteen months there was plenty of time to retarget it at Jupiter after assessing the Pioneer 10 encounter results. During the mission planning phase John Wolfe and Jack Dyer at Ames had conferred with JPL trajectory specialists and identified an opportunity to send Pioneer 11 from Jupiter to Saturn, providing all had gone well with the Pioneer 10 encounter. The maneuver required the spacecraft to pass in front, rather than behind, Jupiter and to be low in the Jovian southern hemisphere so the spacecraft would be deflected upward and pass up and over the ecliptic plane to encounter Saturn on the opposite side of the solar system. One of the beauties of this trajectory was that the spacecraft would pass almost perpendicularly through Jupiter's radiation belts and thereby receive a lower total radiation dose than absorbed by Pioneer 10. In December 1974 Pioneer 11 reached Jupiter and sent back closeup images of the south polar region from an angle not possible with Earth-based telescopes (not even the later orbiting Hubble Space Telescope). The images were rapidly processed and sent out almost in real time on public television.

After its long voyage across the solar system Pioneer 11 approached Saturn in September 1979, right on target after traveling more than one billion miles. Opinions on how the spacecraft should be targeted for its Saturn flyby had differed. I considered the Pioneers ideal "high-risk" vehicles; that is, they were relatively inexpensive and could be sent into potentially perilous regions where we would not dare send the later, more sophisticated and expensive Grand Tour (Voyager) spacecraft. We had already sent the Pioneer spacecraft boldly into outer space to find out if it was possible to pass through the Asteroid Belt, and the craft had braved the radiation hazard of daringly close passes at Jupiter, so I wanted the Pioneer team to identify the passage at Saturn that would promise the greatest scientific return, even at risk to the spacecraft. If that high-payoff trajectory were to pass *inside* the rings, through the gap between the rings and Saturn itself, so be it. It would be productive (and exciting for the public), even if debris in the ring plane should terminate our brave little craft. However, by 1979 I had transferred from NASA Headquarters to join the Director's Office at GSFC, and my successor, the multitalented A. Thomas Young, was quite a bit more conservative. He decided to send Pioneer 11 well outside the ring plane, at the distance where the coming Voyager 2 spacecraft (trailing four years behind) would have to pass if it wanted to continue from Saturn on to Uranus and Neptune.

What troubled me about that "safe" approach was that a sample of one did not give much statistical confidence either way: even if Pioneer 11 failed its passage past the rings there would still have been a great incentive to try again and send Voyager 2 on the trajectory that gave it a chance to make the first flybys of Uranus and Neptune. But all of that is now just speculation overtaken by history. Pioneer 11 did indeed survive its Saturn passage and generated confidence (even if not statistical) that Voyager 2 would also survive, so Tom Young probably made the proper choice. By the time of the Saturn flyby Tom had transferred from Headquarters to Ames as its deputy director. At a postflyby press conference he was pleased to say, "We can report to Voyager: 'Come on through, the rings are clear.'"

Pioneer 11 was able to view Saturn's rings back-lighted, a view not possible from Earth. These views showed considerable scattered light in Cassini's Division, a gap in the rings 6,450 kilometers (4,000 miles) wide first discovered by the astronomer Cassini in 1675. A spacecraft would not be able to pass through the gap. Radiation belts were discovered, although much weaker than Jupiter's, and the rings were found to absorb charged particles, creating a ring-plane zone shielded from radiation. Pioneer 11 discovered a Saturn magnetosphere one thousand times as great as Earth's but only one-twentieth that of Jupiter. The magnetic field was aligned with the axis of rotation, not tilted like Jupiter's and Earth's.

Into the Galaxy

All of the Pioneer spacecraft since Pioneer 6 have had remarkably long lifetimes, a tribute to their builder, TRW. Pioneer 11 functioned until

RTG power gradually dropped below the operating threshold in 1995. Pioneer 10 did even better. In its 4 November 1996 issue, *Time* published a full-page article titled "Still Ticking—A Quarter-Century after Its Launch, Pioneer 10 Is Alive and Calling Home." The plucky spacecraft was still sending useful scientific data from a distance of over six billion miles, twice as far from the Sun as the outermost planet, Pluto. However, the power from its four RTGs had degraded by two-thirds, permitting operation of only two of its instruments. NASA's Wes Huntress, associate administrator for Space Science, decided to officially terminate Pioneer 10 science operations on 31 March 1997. Occasional tracking was still possible, even though Pioneer's eight-watt signal when collected by the seventy-meter antennas of JPL's Deep Space Network (DSN) is only a trillionth of a trillionth of a watt, an energy level so minute it is almost beyond imagination. The RTGs finally were no longer able to power even the transmitter, and in 1998 Pioneer 10 ended its long and highly productive life.

During the extended missions there was some hope that one of the two Pioneers would find the heliopause, the boundary of the heliosphere where the solar wind encounters the interstellar wind. Pioneer 11 is heading in the expected direction of the heliopause whereas Pioneer 10 is going in the opposite direction, down the tail of the heliosphere. Although Pioneer 10 had reached as far as 67 AU by March 1997, the heliopause is currently judged to be located somewhere between 110 and 150 AU, so the Pioneers will have to pass the baton to the faster Voyager 1. The two small spacecraft could not do it all, but they were indeed fantastic pioneers in the exploration of our solar system.

CHAPTER 4

1974 AND 1976

HELIOS 1 AND 2 SOLAR PROBES

As continuing evidence of the relatively easy sell space-race days of the mid-1960s, Helios was an important deep-space mission that we did not have to campaign for—the White House created it for us. In 1966 President Lyndon Johnson and Chancellor Erhard of the Federal Republic of Germany (West Germany) met in Washington to forge closer ties between their two nations. The Germans appreciated better than many Americans that the Apollo program was not just an effort to recapture international prestige from the Soviet Union and its spectacular Sputniks but also a great impetus to technological advancement. Achieving the reliability necessary for safely putting American astronauts into space, for example, required much tighter quality control in the manufacture of solid-state electrical devices and the development of rugged solid-state computers with greatly increased capability. Erhard believed that undertaking a technically challenging spacecraft project would provide a similar stimulus to German technology and a resulting boost in Germany's exports and economy. Lyndon Johnson, as leader of the Senate and, later, in the White House as vice president and then president, had played a major role in building the U.S. space program. He and Erhard readily agreed that a joint German-U.S. space project would be a mutually beneficial endeavor.

Seeking a Challenge

As their object of scientific study the Germans chose the Sun. A number of their scientists already were engaged in solar physics, and the Sun is certainly all-important to Earth and its inhabitants. Consider Earth's climate alone. During the second half of the seventeenth century the historians of the developed world—that is, Europe—recorded a miniature ice age, including heavy snow in normally mild England. Even at that time excellent records were kept of dark spots (storms) on the Sun, which could be easily imaged and traced by letting sunlight pass through a pinhole and onto a sheet of paper. The miniature ice age corresponded exactly with an almost complete disappearance of sunspots. It was determined and later verified that the variations in solar activity have a significant impact on Earth's climate—and on many other terrestrial functions.

In the twentieth century we are affected almost daily by the rise and fall of solar storms, which interfere with radio communications. Better understanding the emanations from the Sun could be beneficial to everyone, the Germans reasoned, and hence a solar probe mission seemed like the ideal scientific choice for a space mission. To make the mission especially challenging, the Germans proposed to send a spacecraft over two-thirds of the way to the Sun, to .3 AU, exposing the spacecraft to solar heating eleven times that experienced at Earth's orbit. Designing a spacecraft to survive that severe environment proved to be most challenging indeed.

U.S.-German negotiations got underway and teams were assembled. NASA would contribute two Atlas-Centaur launches from Cape Canaveral, tracking by the worldwide DSN and use of JPL's Space Flight Operations Facility during the early stages of the mission. In return U.S. scientists would be invited to be part of the science team and would get to fly their instruments on this pioneering mission. When it

came time to name the project, the Germans turned to Greek mythology, borrowing the name of Helios, the god of light and warmth,
usually pictured racing across the heavens in a chariot drawn by fiery
stallions.

Getting Organized

By 1969 the Helios Project was well organized and moving forward.
At NASA Headquarters the deep-space mission was assigned to Code
SL, where Don Hearth appointed Glenn Reiff program manager (later
replaced by Fred Kochendorfer when Glenn left NASA) and negotiated
with Code SP to have the very able and gentlemanly Albert Opp assigned as program scientist. As U.S. project manager, NASA appointed
Gilbert Ousley from the Goddard Space Flight Center (GSFC). With
years of experience in Europe, stationed in Paris as NASA's European
representative, Gil was a natural choice. He was not only a good manager but also personable, enthusiastic, and great in dealing with people.

Gil viewed his job on Helios as getting the best possible support for
his German colleagues. They loved him right from the start. One thing
that always impressed me about Gil was how he could conduct dinner
parties in Paris and never touch the wine, and join heartily in loud beer
parties in Germany while sipping only apple juice. It surprised me that
I never heard him even attempt to speak French or German, though I
know he must have had a pretty good vocabulary in both those languages. The only German I recall him speaking was "Apfelsaft, bitte"
when ordering his apple juice. Yet the Europeans loved him. And he
served them well.

Gil recruited James Trainor, a respected and experienced physicist at
Goddard, as his project scientist. Jim was a member of the Cosmic Ray
team on Pioneers 10 and 11 and was selected to lead the galactic and
solar cosmic ray experiment on Helios. Another essential member of
the Helios management team was Kurt Heftman, who arranged for all
of the project-support activities at JPL.

It was planned that each of the two Helios spacecraft would carry
ten instruments to measure solar wind (protons, alpha particles, elec-

trons), magnetic field, electric field strength, cosmic radiation, gamma rays, X-rays, zodiacal light, and dust particles. In addition to Jim Trainor, a number of other Goddard scientists participated in these experiments, including Norm Ness, Len Burlaga, Susan Kayser, Robert Stone, Michelle van Hollebeke, Nand Lal, and Frank McDonald. Also collaborating were D. A. Gurnett and R. R. Anderson from the University of Iowa, Paul Kellogg from the University of Minnesota, and G. S. Levy, C. T. Stelzried, and B. L. Seidel from JPL. As about half the science team was from the United States, the U.S. space program was going to benefit substantially from this cooperative international project.

In Bonn the Helios Project was administered by the Federal Ministry for Research and Technology (Bundesminister für Forschung und Technologie, or BMFT), which was represented by its Helios program manager, the always friendly and cooperative Karl Kaesmeier. The prime contractor for the project was MBB-ERNO, headquartered in Ottobrun, just south of Munich. Mission planning and daily operations starting with Helios 2 were performed at the German Space Operations Center (GSOC) at DFVLR in Oberpfaffenhofen, a few kilometers west of Munich.

The clear German leader of the Helios Project was Ants Kutzer, the Helios project manager from DFVLR. Ants was a large man with a forceful personality and reminded me a great deal of Jim Martin, our strong project manager on Viking. Both men were prematurely white-haired, and there was never any doubt with either of them who was leading the troops. Senior physicist Herbert Porsche was appointed Helios project scientist for the German team, although Kutzer did not seem to lean on him for much scientific leadership. A key manager, who was to grow in stature as the project progressed, was Martin Schurer, who headed the operations effort at DFVLR in Oberpfaffenhofen.

The first major spacecraft design review was held at the MBB plant in Ottobrun in September 1970. All proceedings were conducted in English. To my disappointment, the American contributors had made no great effort to learn German. There were exceptions, of course, such as Jim Trainor, who practically lived in Munich in his active roles as U.S. project scientist and cosmic ray experimenter. The Germans took a different tack. First of all, they adopted all of the American terms for spacecraft components and subsystems. They said that we had nice short, precise words for every spacecraft part and function, whereas German words, which had to be made up, tended to be descriptive (somewhat like Native American words) and hence unduly long. They liked the short words better. Having made that decision, they decided to write all of their design manuals in English. The next logical step was to conduct all of their meetings in English. All their people had had language classes in English in school and most welcomed the opportunity to become more fluent. By the time we launched the Helios spacecraft all the German team members were speaking English beautifully. We lazy Americans could only get by with ordering our "bier" and "apfelsaft."

To obtain the best independent critique of the Helios spacecraft design, Gil Ousley had recruited the services of Herman Legow, a man who spoke with the quiet authority that came from nearly twenty years' experience in designing and developing the stream of small- and medium-sized spacecraft built at the Goddard Space Flight Center. Legow had brought with him some of the most experienced engineers from Goddard, such as the world's leading expert in spacecraft stabilization, Henry Hoffman. In September 1970, guided by Legow and Kutzer, the Helios design-review meeting ran smoothly and dug thoroughly into design details. After several days, the participants were pretty well drained.

Our German comrades had cleverly timed the design review to coincide with Octoberfest in Munich. At the conclusion of our last meeting we were all to gather at tables reserved by MBB in one of the huge Octoberfest pavilions. Our meeting ran overtime, and when we arrived at the pavilion, almost two hours late, we were advised that they had held the tables for an hour and then given them away. We were faced with a noisy crowd of three thousand people, jammed tight at every table. My wife Anne, who had accompanied me on the trip, was dismayed and wanted to return to our hotel, but up stepped the dynamic Martin Schurer of DFVLR, with a big smile on his face, holding a long-stemmed red rose, which he handed with a flourish to Anne. "Follow me," he urged. As we wandered through the closely packed tables he persuaded one group to squeeze together even tighter to make room for Anne and me. "Have fun," he said, as he departed to help others of our group.

Surprisingly, we did. I spoke only a few words of German, and nobody else at the table spoke much English, but before long we knew where everyone came from, who was married, how many children they had, and so on. The one-liter mugs of beer, which went with the traditional whole roast chickens, started arriving. We were starving and dug right in. Now, Anne and I are very moderate drinkers, maybe two glasses of wine at a long party. Almost never any beer. But that night we bought seven liters of beer! Curiously, we felt perfectly fine the next morning. Unbelievable? Well, they tell us that the Octoberfest beer is the first of the season out of the big vats and is very light, going down ever so smoothly, almost like water. Also, I am sure Anne and I did not consume the entire seven liters by ourselves; we were sharing all around, everyone pouring into any empty mug so everyone always had some to drink as the oompah band played rousing drinking songs. Anyway, within half an hour we felt we were with good friends, arms around one another, singing songs. We had a great time and learned the friendly warmth the Germans call Gemütlichkeit. Somehow that evening seemed to set a welcoming tone for the cooperation to come on the Helios Project.

Martin Schurer turned out to be a real "take charge" guy. Somewhat domineering but highly likeable. He was a bold skier and a world champion glider pilot after World War II. His lovely wife, Roswitha, is certainly one of the most charming persons I have ever met. Martin and Roswitha took over the roles of official host and hostess, which meant not only dinner at their beautiful new home in Oberpfaffenhofen, a quaint village a few kilometers west of Munich, but getting us perfect third-row seats at the Passion Play in Oberammergau. A marvelous experience, even if you don't fully agree with their biblical interpretation. The play is put on only every ten years, and we were fortunate to be there in 1970.

Clever Solutions

The design review had highlighted the difficulty of the mission the Germans had selected. The Helios spacecraft would spin, with its spin axis perpendicular to the ecliptic plane, so that in half of every revolution

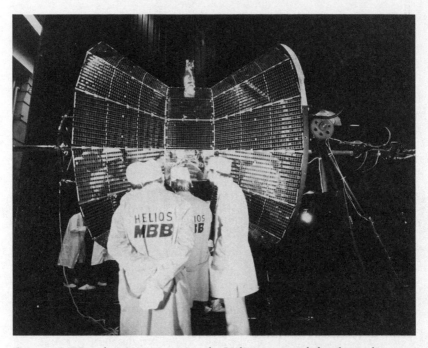

German MBB technicians preparing the Helios spacecraft for thermal-vacuum testing at JPL to an intensity of eleven Suns. The surface is coated with second-surface mirrors in order to reflect the extreme solar radiation. Helios 1 was launched on 10 December 1974 and Helios 2 on 15 January 1976, penetrating to .29 AU from the Sun. (Courtesy NASA)

it could radiate heat away into cold space. The craft was given a unique spool shape with flared ends to maximize the end surface area radiating to space, but there was no way it could fully dissipate the intense heating of eleven solar constants. Most of the solar intensity would have to be reflected off the craft. Simple but ingenious second-surface mirrors developed at NASA Ames provided the solution. Glass plates like those used to cover silicon-based solar cells were silver coated on their back side and used on all Helios spacecraft surfaces not covered with active solar cells. The silvered second surface provided almost 100 percent reflectivity, and the external glass surface radiated heat with a very high emissivity. The combination worked beautifully. To complete the exterior surface, high-temperature solar cells were developed and attached to the craft with newly developed special adhesives. These so-lutions were clever, but the resulting weight of the spacecraft was be-ginning to strain the capability of the NASA-provided Atlas-Centaur launch vehicle.

By 1970 Code SL had settled its forthcoming Viking Mars mission on an orbiter and soft lander design that would require the develop-ment of a new high-energy Titan 3–Centaur launch vehicle configura-tion. With the Viking Project projected to cost the better part of a bil-lion dollars, we were getting a bit nervous about it going on the very first launch of a new vehicle configuration. We offered a deal to the Germans: NASA would replace the two Atlas-Centaurs for Helios with two of the more expensive Titan 3–Centaurs. The increased payload capability would save them the costly struggle of trying to reduce the weight of their Helios spacecraft. We would launch them on schedule in 1974, a year ahead of Viking. There would be a somewhat greater risk in going on the first Titan 3–Centaur launch, but we would then give them a second launch in 1976, a year after Viking, when any bugs with the launch vehicle should be fully resolved. They liked it. We had a firm project plan that everyone was happy with.

The prototype spacecraft was completed on schedule, functionally tested at MBB, and subjected to simulated launch loads on the mam-moth centrifuge at NASA Goddard. The prototype was then shipped in the spring of 1974 to JPL in California for simulated space envi-ronmental testing. The large thermo-vacuum tank at JPL, called the Large Solar Simulation Facility, could pump down to a hard vacuum and had a focused array of mercury and xenon lamps that could radi-ate the spacecraft with simulated sunlight at the equivalent intensity of twelve solar constants, higher than expected in flight at .3 AU. Test re-sults were close to expectations, and the go-ahead was given to com-plete assembly of the flight spacecraft.

Into the Fire

On 10 December 1974 NASA gave Helios 1 a flawless launch at Cape Canaveral. Counting drop-off solid-propellant boosters and the two stages of the Titan, the Titan 3E–Centaur–TR-364-4 was a five-stage launch vehicle—very potent indeed. The Germans were delighted, and so were our Viking Project people. Tended by the operations team at the SFOF in Pasadena and a crew in training at the GSOC in Oberpfaffenhofen, Helios 1 reached its perihelion (closest approach to the Sun) of .31 AU on 15 March 1975. Heating was intense, but all systems functioned as planned, except for the failure to deploy of one of two antennas for Gurnett's plasma wave experiment. Gurnett dismissed this as "not really very serious." The Helios mission could already be declared a great success, not only for its science return but also as a technical triumph for Germany.

Based on Helios 1's orbital performance some minor modifications were made on the Helios 2 spacecraft, mainly to better isolate the viewing of various experiments. During 1975 Titan 3–Centaurs successfully launched the two Viking spacecraft on their way to Mars. This fine launch vehicle then, on 15 January 1976, after a perfect countdown at the cape, propelled Helios 2 swiftly on its way to pass even closer to the Sun than its sister Helios 1. A perihelion of .29 AU was reached by Helios 2 on 17 April 1976 at the fantastic speed of 252,000 kilometers per hour. Solar heat flux was at 11.86 solar constants, but Helios 2 survived and returned excellent data. Control of both Helios spacecraft was now from the GSOC in Oberpfaffenhofen.

On Helios 2 the thermal design of the spin thrusters was modified to improve performance. Any change is a gamble, and this one did not pay off: spin thruster temperatures ran far above Helios 1 values. Still, Helios 2 returned for eight perihelion passages before failing on 3 March 1980. In 1984, ten years after launch and after nineteen perihelion passages, Helios 1 was still functioning and returning valuable data. The Germans celebrated the anniversary with several days of ceremonies and a scientific symposium in Munich. They invited the American team members and paid all our expenses. Very nice of them and most enjoyable.

The anniversary symposium summarized all the scientific findings produced by the Helios Project. U.S. program scientist Al Opp summarized: "The direct propagation of solar disturbances and shock waves through interplanetary space have given new insight into the structure of the interplanetary medium, the generation of the solar wind at the Sun, as well as fundamental new knowledge of tenuous, collisionless plasmas. The fact that the Helios spacecraft operated from solar minimum into and through solar maximum, has given the scientific world a detailed, close-in view of the Sun over vastly differing solar conditions. It has enabled scientists to observe cosmic rays coming into the solar system from our galaxy, and when combined with results from deep space probes and Earth orbiting satellites, has given a detailed picture of the structure of the solar system and of the characteristics of low energy galactic cosmic rays."[1]

Helios became a model for joint international space projects to come. The scientific results of the project continue to be valuable, as do the many overseas friendships that endure to this day. Perhaps here was the greatest value of all. As noted by Ants Kutzer in the tenth-anniversary publication, "The appreciation of commonality of purpose added a richness to the cultural interchange over and above the normal pride of completion of a challenging and complex technical project such as Helios. The participants in Helios project are proud to share such experiences gained and deep personal friendships developed."

Gil Ousley and his two trusted deputies, Charlie White and Bill Witt, went on to manage a number of other joint U.S.-European space projects, which kept the well-liked trio traveling across the Atlantic, where they continued in close touch with their many friends from Helios.

1973

MARINER 10 VENUS AND
MERCURY FLYBYS

Selling the 1971 Mariner 9 mission to Mars and the 1972–73 Pio-neer 10 and 11 missions to Jupiter had been relatively easy, and the Helios missions to the Sun had been handed to us as a presidential gift of international diplomacy. The Sun is of clear importance to the human race, and Mars and Jupiter are both spectacular planets with scientific significance as well as popular appeal, so missions to study them were easy sells. If we were going to better understand our solar system, however, we were going to have to learn a lot more about *all* the planets.

The National Academy of Sciences had repeatedly reminded NASA that in planetary exploration there should be three major goals: "To further our understanding of: (1) the origin and evolution of the solar system, (2) the origin and evolution of life, and (3) Earth by comparative studies of the other planets." To progress toward the third goal we especially needed to sell missions to our neighboring terrestrial[1] planets Mercury, Venus, and Mars. We had a good start on Mars (Mariners 4, 6, 7, and 9) and at least a beginning on Venus (Mariners 2 and 5), but we knew that selling a mission to Mercury was going to be diffi-cult. Space scientists would see the need, but how about White House budgeteers and congressmen? All we could tell them about Mercury was that it was an airless, arid body, not too much larger than our Moon. It was not going to be an easy sell under the most favorable of circumstances, and we were sailing into decidedly unfavorable weather. By 1968 it was growing more and more difficult to get new space science missions into the NASA budget. The Apollo Project was consuming a great deal of money, as was the escalating war in Vietnam, the NASA budget was coming down from its peak year of 1965, and both the administration and Congress were increasingly resisting any additions to the budget. Clearly, getting a mission to Mercury into the NASA budget was not going to be easy. No way. In fact, nothing about the Mercury mission from beginning to end would come easily.

Four Hurdles to Clear

In this tougher era, getting any new planetary mission into the budget required clearing a series of four hurdles in sequence: (1) endorsement by the scientific community as represented by the National Academy of Sciences and its Space Science Board (later known as the Space Studies Board), (2) convincing NASA top management to include our new planetary mission in the NASA budget request to the president rather than one of the many other promising candidate missions vying for a new start, (3) getting the Bureau of the Budget (later the Office of Management and Budget) to leave our new start in while it was cutting the NASA budget request down to something the president was willing to sign, and (4) getting congressional approval in the final authorization and appropriations bills. For little known Mercury these hurdles were an especially daunting challenge; the Headquarters Code SL team knew it would need something "special," sound arguments and some gimmick, to sell this one.

Don Hearth, Don Rea, and I worked very well together as the "marketing" team within Code SL. Among the many interesting candidates for planetary missions, we looked for missions that had special strengths in getting through each of the four gates. Don Hearth and I engaged in long and vigorous debates over the merits of new-start candidates. I probably gave too much weight to technical factors, whereas Don mixed in a lot more political pragmatism, yet we invariably came to the same conclusion. One of the factors we both looked for was uniqueness in launch opportunities. Something to counter the question "Why now?" During the battles for priority within NASA it was all too easy to be put off "until next year," and this slip could go on year after year. In 1973 we found that we had something special—an especially favorable launch year in which to utilize "gravity assist" for a multiple-planet mission to both Venus and Mercury.

Celestial Billiards

Gravity assist is a technique that takes advantage of the gravity pull on a spacecraft as it flies past a planet. The planets are all traveling at high speed in their orbits around the Sun. If the spacecraft flies behind a planet it is pulled in the direction of that planet's motion around the Sun and hence the spacecraft picks up speed. If the spacecraft flies in front of the planet it is pulled back from the direction of that planet's motion around the Sun and the spacecraft loses speed. Jupiter, being by far the most massive of our planets, can give the most dramatic gravity assist. For example, Voyagers 1 and 2 would later pass behind Jupiter and pick up enough velocity to reach Saturn, and the European Ulysses spacecraft would pass in front of Jupiter and both deflect itself out of the ecliptic plane (the plane around the Sun roughly containing the or-bits of all the planets but Pluto) and kill enough velocity to drop in-ward toward the Sun. Pioneer 11 would perform an even more com-plex flyby near the south pole of Jupiter, pass over the ecliptic plane, and reach Saturn on the opposite side of the solar system.

There is disagreement over who first conceived of gravity-assist tra-jectories. Arthur C. Clarke, for example, credits Derek F. Lawden, who in September 1954 published a paper on "Perturbation Manoeuvres" (*Journal of the British Interplanetary Society* 13, no. 5), but others claim the honor. Whoever the inventor is, Dr. Homer Joe Stewart of Caltech and JPL should get credit for clarifying the concept. When I was a grad student at Caltech in 1950 Homer Joe was known as the most brilliant student to ever pass through the graduate school, even though he was reported to have spent more time shooting pool than studying or doing research. Homer reasoned that a spacecraft flying by a planet was like the elastic collision between two billiard balls, although in space flybys the balls are of vastly different mass. At flyby (gravitational attraction providing the equivalent of the collision) the spacecraft can pick up almost the entire speed of the planet while the planet slows down by an equal amount of momentum (the product of mass times velocity), but due to its enormous mass, the planet's change in velocity is negligible. Voyager 2 was later going to play this pool game to the hilt in its Grand Tour of the outer planets.

Two young grad student summer employees at JPL, Michael Minovich and Gary Flandro, took this celestial billiards concept and developed the mathematics and trajectories that proved the validity and potential of the gravity-assist concept. Minovich identified the Venus-Mercury opportunity; Flandro concentrated on the outer planets. Flandro believes that much of the credit for developing the analytical techniques for calculating gravity-assist trajectories should go to the prolific Krafft Ehricke and his textbook *Space Flight* (vol. 2, *Dynamics*), published in 1962. Our proposed 1973 Mariner 10 Venus-Mercury mission (MVM 73) would be the first to employ gravity assist to fly by two planets, Venus and Mercury, with one spacecraft. (In 1974 Michael Minovich was invited to JPL for Mariner 10's Mercury encounter and recognized for his contribution to this first gravity-assist mission.)

Mercury is more than half way toward the Sun from Earth, and getting there is not easy. It would take a large and expensive Titan 3C–Centaur launch vehicle to send a modest-sized Mariner spacecraft directly to Mercury. But a factor that countered any move to slip the new start to "next year" was that 1973 was a most favorable year (the best in fifteen years) for a Venus gravity assist to slow a spacecraft and deflect it in to a Mercury flyby. If launched in 1973, the mission could be accomplished with just a proven Atlas-Centaur. In addition we believed that we had good scientific justification to learn more about Venus and Mercury. Hearth, Rea, and I agreed that we should go for it.

Over the Hurdles

Getting past hurdle number one, science endorsement, did not appear too difficult. Venus, in many ways similar to Earth, had somehow evolved a very dense carbon dioxide atmosphere that had trapped outgoing infrared radiation (the greenhouse effect) and raised its average surface temperature to 730 degrees Kelvin (850 degrees Fahrenheit), enough to melt lead. It could tell us a great deal about where Earth was heading as humans continue to burn fossil fuels and dump more and more carbon dioxide into our own atmosphere. Moreover, even after Mariners 2 and 5 and several Russian Venera probes, no one had yet taken closeup photos of Venus. Little could be seen from Earth-based telescopes, but there was evidence that cloud structure would be seen in the ultraviolet wavelengths. Mercury, on the other hand, probably never had a significant atmosphere, so that its surface promised to give a clear picture of impact history during the early stages of planet formation in our solar system.

We first tried out our scientific justification on our newly formed advisory group, the Lunar and Planetary Missions Board, whose members were respected scientists covering all aspects of planetary science: John W. Findlay (chairman), James R. Arnold, Al F. Donovan, Von Eshleman, Tommy Gold, Clark Goodman, John S. Hall, Harry H. Hess, Francis S. Johnson, Josh Lederberg, Lester Lees, Gordon J. F. MacDonald, Michael McElroy, George Pimentel, C. S. Pittendrigh, Frank Press, Eugene M. Shoemaker, James Van Allen, and Wolf Vishniac.

During his leadership of Code SL Oran Nicks had of necessity concentrated on engineering technology development. The first steps had to be to get launch vehicles and spacecraft to work; establishing good relations with the scientific community was not a top priority. The LPMB was created to increase scientific involvement in planetary programs, and Don Hearth welcomed useful input from the board. Board members responded with enthusiasm and helpful suggestions to our presentations on Venus-Mercury opportunities. They were a fine group to work with and very constructive. John Naugle credits them with getting planetary scientists to stop pushing just their individual technical specialties and pull together for specific planetary projects. With favorable support from the influential LPMB members, the Space Science Board endorsed the proposed 1973 Venus-Mercury mission in June 1968. We had cleared hurdle number one.

Hurdle number two, the competition within NASA, has always been the toughest to clear. You have dedicated people within the various program offices at NASA Headquarters with important charters and objectives to accomplish and all with promising and potentially rewarding new-start projects. Each space science group has at least one new-start candidate they have cleared through hurdle number one, with influential scientists applying pressure on NASA top management to approve their favorite new project.

We were having a tough time in this fiercely competitive environment. For one thing, we were competing with ourselves in that we were also proposing to advance our exploration of Mars through a new project called Titan Mars '73 (see chapter 6), which would hard land instruments on the surface of Mars. In a planetary priorities race Mars was going to win out over either Venus or Mercury. Dr. John Naugle, NASA's associate administrator for Space Science and Applications, liked the science potential of the Venus-Mercury mission and appreciated the unique launch year of 1973 but could see no way to fit the estimated $140 million cost of the project into NASA's science budget.

Then out of the blue came a remarkable letter from Dr. William Pickering, director of JPL. Absolutely and flatly, with no qualifiers, Pickering guaranteed that JPL would complete the mission (one launch only) for $98 million, including data analysis. I did not really believe it could be done for that amount, but the letter had great appeal to John Naugle and Homer Newell. They had always had a difficult time working with Pickering, who clearly believed that he reported directly to the NASA administrator, even though according to NASA's organization charts JPL reported to Naugle. Here was a chance to hold Pickering to a firm deal with no appeal to the NASA administrator for later budget relief. I referred to the Pickering guarantee as "the letter signed in blood," and it turned out to be the deciding factor in getting the 1973 Mariner 10 Venus-Mercury Project into the NASA budget request for fiscal year 1970.

The next hurdle was the White House, or more immediately its Bu-

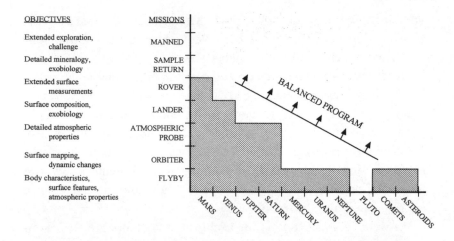

Chart developed by Robert Kraemer and used by NASA Headquarters to illustrate its Balanced Strategy for the exploration of the solar system. Initially employed in 1969 to illustrate the NASA-proposed projects for start in fiscal year 1970, the chart was utilized in many forms over the years and is shown here updated to include missions launched through 1999, including international missions (for example, Russian Venera landers on Venus and Russian-Japanese-European flyby missions to Comet Halley).

reau of the Budget. We knew that BOB liked orderly planning. With so many worthy candidates for NASA's new starts, and with the various program office people pushing their own favorites, BOB had a difficult time determining which missions were well planned and worthy of early funding. To illustrate that Venus and Mercury missions were an integral part of our planned exploration of the solar system I devised a planetary strategy chart. Across the bottom of the chart I listed the bodies in the solar system in (somewhat arbitrary) order of scientific importance and accessibility. Ascending the vertical axis of the chart are the advancing stages of exploratory missions, ranging from simple flybys to probes and landers and eventual sample return missions. Two possible strategies were to (1) concentrate on just the top-priority planet, Mars, and pursue missions all the way to sample returns to Earth, or (2) concentrate initially on simple flyby reconnaissance missions to all the classes of bodies in the solar system, including representative comets and asteroids. With a diagonal line on this chart I indicated the "Balanced Strategy" that we advocated, which would press forward with a mix of both reconnaissance and in-depth exploration at whatever pace could be afforded by the NASA budget.

As I explained in a position paper I wrote in November 1968 on the "Objectives of Planetary Exploration," we believed that the Balanced Strategy was supported by an NAS study, *Planetary Exploration: 1968–1975*, released in June 1968. That report backed up our objective of studying all the planets by citing the contribution that the study of Mars' atmosphere had made to recognizing the importance of radiative heat transfer in predicting weather on Earth. To quote the NAS report, "In the case of the atmosphere of Mars, there is thus a direct and demonstrable connection between study of that atmosphere and a better understanding of our own. From this better understanding, we can perhaps expect advances in the techniques of long-range weather forecasting, and eventually the development of effective methods for weather modification. The significance of the exploration of the Martian atmosphere to studies of the Earth is by no means unique. We can expect further advancement in the understanding of our own planet as a result of the study of the other planets." Of course, after Mars, the "other planets" most like Earth are Venus and Mercury, the objectives of our proposed new mission.

It amazed me how effective my rather simple-minded Balanced Strategy chart proved to be. My initial version showed how far we could proceed with our proposed program for fiscal year 1970. I showed it to the LPMB and they were pleased with the balanced approach. On 11 July 1969 I featured it in a formal presentation that Don Hearth, Don Rea, Bill Brunk, and I put on for Dr. Thomas Paine, the NASA administrator under Lyndon Johnson, and it went over well. For Dr. Paine I added another version of the chart showing the progress we could make through the 1970s. John Naugle tells me that he thought my charts were "just great, because they showed a logical planetary program for the next thirty years." My chart was especially effective at BOB, where they were comforted by it as a demonstration that we had really thought out our mission priorities. I used this chart for years thereafter, coloring in the new missions as we had them approved. They always fit in nicely with the Balanced Strategy line and continued to reassure BOB and its successor OMB. In fact, OMB was thereafter the easiest of our four hurdles.

Finally, having cleared the BOB hurdle, we had to get congressional approval. I thought this would be easy, as planetary missions had always benefited from good support in the congressional committees, but it would not prove so simple for Venus-Mercury. Our toughest and most influential congressional committee at that time was the House Authorization Subcommittee for Space Science and Applications chaired by Rep. Joseph Karth of Minnesota, who was relatively young for a congressional chairman and still full of athletic energy and endurance. I would rate him one of the most intelligent and capable members of the House, and in no time he became quite knowledgeable about NASA's programs. He was also fortunate to have an outstandingly capable counsel in the personable Frank Hamill. Between the twosome of Karth and Hamill you faced a stream of probing and insightful questions when you testified at one of Karth's hearings. We spent a lot of time doing our homework before going before Karth.

Space Science did well when the multitalented Ed Cortright was the lead witness for NASA's Space Science and Applications programs (*applications* meant applying space techniques to Earth services, such as weather satellites, Earth resources satellites, communications satellites, etc.). Ed played golf with Karth, they became good friends, and Ed was also very smooth in hearings. When John Naugle took over as associate administrator for Space Science and Applications things got rougher with Karth, who immediately assumed that Naugle as a scientist would favor science projects over Earth application projects. That was really not true, as John was extremely conscientious about pursuing both aspects of his charter, but he was never able to convince Karth that he was not biased.

Let me digress to tell you more about John Naugle. First of all, he is one of my favorite people. Just why may not be immediately obvious. Rated by some management consultant, he would not score especially well. He does not possess a dominating personality. Of no more than medium height, his outstanding feature is his trademark bushy, mutton-chop sideburns. He speaks rather deliberately, giving some the (false) impression that he is not all that sharp. He would flunk out in

TOP: A southern region of Jupiter as viewed by Voyager 2 on 3 July 1979. Although the huge cyclonic storm know as the Giant Red Spot has lasted for many years, other features in the Jovian clouds change rapidly; for example, the white spot below the Giant Red Spot was not there during the earlier passage of Voyager 1. (Courtesy NASA) BOTTOM: Early view of the rock-strewn Martian surface in the Chryse plains area as seen by the facsimile cameras of the Viking 1 Lander in late July 1976. The sky is not the blue characteristic of Earth. (Courtesy NASA)

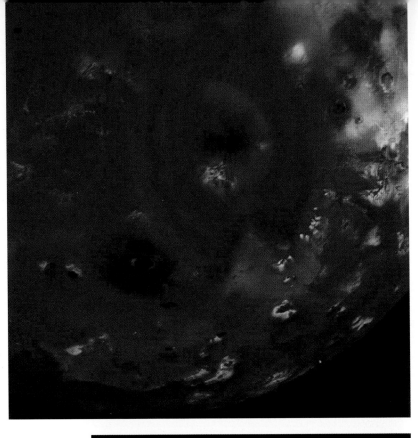

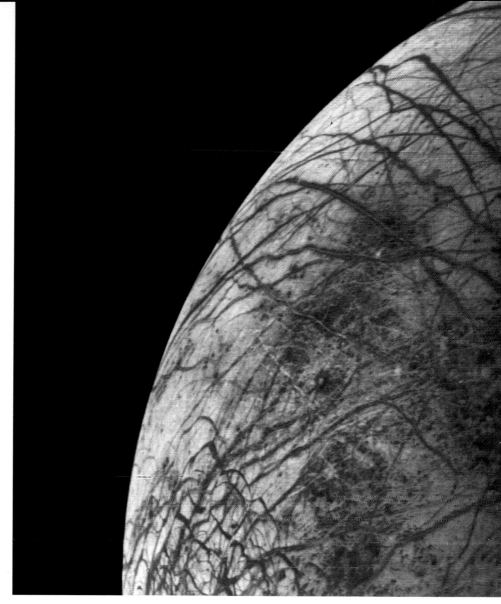

ABOVE: The icy surface of Europa (color enhanced), a satellite of Jupiter, as viewed by Voyager 2. Europa is the smoothest satellite ever observed, albeit fractured, with no mountains or deep valleys or craters, indicating a surface of recently frozen ice and an underlying ocean. (Courtesy NASA) BOTTOM OPPOSITE: A multicolored Voyager image of Io (a satellite of Jupiter) that was immediately likened to a "pizza." Scientists were intrigued by the colors and then became really excited when eight active volcanoes, spewing out colorful sulphur and sulphur dioxide, were discovered. (Courtesy NASA) TOP OPPOSITE: A large active volcano on Io named Pele. The red color (somewhat enhanced here) identified it immediately as "the pepperoni on the pizza." (Courtesy NASA)

Saturn as viewed by Voyager. (Courtesy NASA)

The rings of Saturn as viewed by Voyager 2 and color enhanced to make more visible what are apparently variations in chemical composition in the different parts of the intricate ring system. (Courtesy NASA)

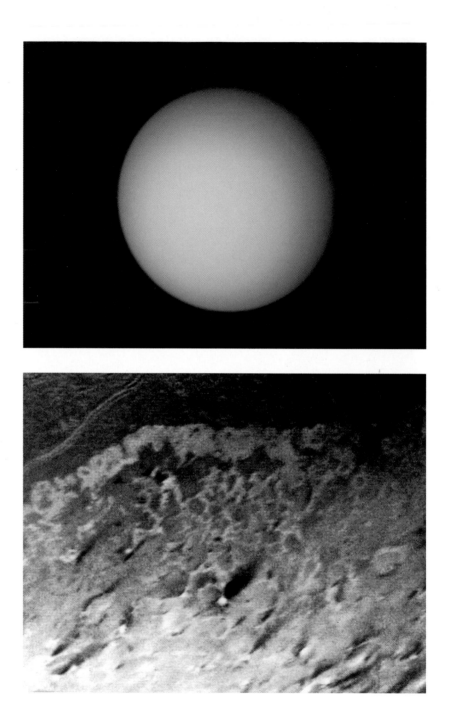

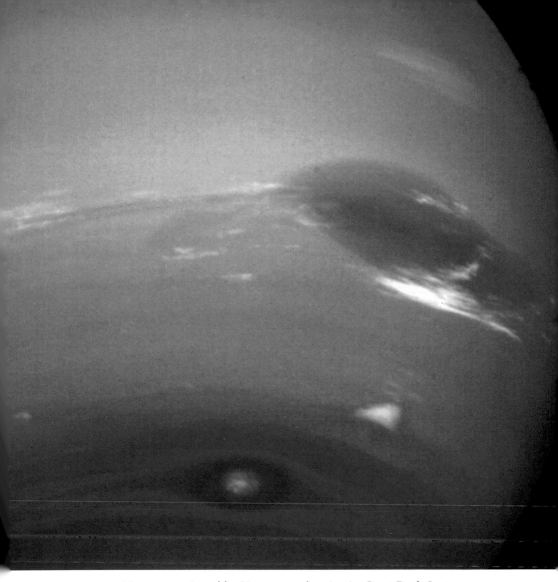

ABOVE: Neptune as viewed by Voyager 2, showing its Great Dark Spot.
The smaller white and dark cloud features move eastward around the
planet at different velocities and so change their orientation with respect
to one another. (Courtesy NASA) TOP OPPOSITE: Uranus as viewed
by Voyager 2. With its spin axis uniquely pointed at the Sun, it was a surprise not to see more structure in the atmosphere. However, the wildly
skewed magnetic field was of great interest. (Courtesy NASA) BOTTOM
OPPOSITE: Close-up of Triton (the large satellite of Neptune) as viewed
by Voyager 2, showing varied textures in its surface of frozen nitrogen and
methane. Brown and dark shades among the white and pink indicate the
presence of organic compounds, and the windblown dark plumes are evidence of geysers of organic material. (Courtesy NASA)

A collage of images of the eight planets viewed close-up by spacecraft launched in the golden era of planetary exploration, starting with tiny cratered Mercury (*top*) and working outward from the Sun to yellow Venus, blue Earth, reddish Mars, multihued Jupiter and Saturn, greenish blue Uranus, and blue Neptune. (Courtesy NASA)

the popular management school of Managing-by-Walking-Around (in which I am a firm believer and practitioner). In fact, in the eight years I worked for John at NASA Headquarters he never once visited my office area, even though it was just down the hall. Moreover, John made it clear that he really did not want to know all that I was doing to sell my programs—if I were ever accused of doing too much lobbying he was going to express shocked surprise. Once I got used to this strange arrangement of not keeping the boss fully informed, I liked it. The other program directors and I promoted our programs as best we could, with few restraints, and then John sat as judge of whose program would survive.

My principal competitor for project new starts was the dynamic and colorful Jesse Mitchell, director of Space Physics (later called Physics and Astronomy) Programs. Jesse was a marvelous salesman. He could have done well with used cars. He managed to sell a continuing constant level-of-effort annual budget for small missions called Explorers, getting around the need for new-start approvals, and would start far more of these small Explorer projects than he could possibly fund, getting a foot in the door and, not being tied to any specific launch dates as we had for planetary missions, eventually getting them all launched. Not especially cost effective to drag out the projects, but productive nevertheless. When he later went for a blockbuster project called the Large Space Telescope (later named Hubble) he had the audacity to propose with a straight face that it was just another Explorer mission and therefore did not need a specific new-start approval. As I said, Jesse could have done well with used cars.

With his space physics background, John Naugle would have been expected to favor Jesse's proposed new projects over the planetary new starts I was proposing. I learned that I need not worry about that. Jesse and I both concluded that John would be a fair judge. He had an excellent feel for scientific priorities and was a man of absolute integrity. Every judgment that John Naugle made was intended to advance the best and most productive space science program. Regardless of whether your proposals got past John, you knew he was a fair judge and you had to respect the man.

But Congressman Joe Karth did not share that respect for Naugle.

At a hearing before his committee he would fire a question at John, who would start into a careful, deliberate, and thorough answer, then cut him off at about the third sentence and fire another zinger. It was brutal. At the hearing before Karth on the proposed NASA fiscal year 1970 budget we did our best to present the arguments for Venus-Mercury, and I made sure that counsel Frank Hamill was fully informed, but when Karth's budget markup was released he had deleted Venus-Mercury from the NASA budget. In his report Karth put it in writing that he really believed that Venus-Mercury was a valuable mission worthy of funding but did not believe that Naugle was trying hard enough to advance Earth applications projects and that the subcommittee would consider restoring Venus-Mercury to the budget when he saw evidence of a stronger applications response from Naugle.

We were then faced with getting Venus-Mercury approved by the Senate Authorization Committee chaired by Sen. Clinton Anderson of New Mexico so that we could try to get the project considered by a House-Senate conference committee (in those days the Authorization committees had all the power on individual projects, with the Appropriations committees passing on total funding but not trying to dictate individual projects). Senator Anderson responded well to the White House on most matters, so we and JPL arranged to have Dr. Lee DuBridge, the president's science advisor, testify to Anderson's committee. DuBridge had just come from his job as president of Caltech, which runs JPL for NASA, so he was fully informed and enthusiastic about our proposed Venus-Mercury mission. He did an excellent job before the Senate committee, and the committee approved Venus-Mercury in their budget markup. All went well in the conference committee (remember, Joe Karth really approved of the mission but was just holding it hostage), and Venus-Mercury was solidly approved in the NASA budget. I did not really believe Bill Pickering's pledge of $98 million, so I told the congressional committees that the cost would be in the range of $105 to $115 million, still well below the original estimate of $140 million. The proposed new start had now cleared its fourth and final hurdle and our 1973 Mariner Venus-Mercury mission was solidly in the fiscal year 1970 budget for NASA.

Forging a Team

In Code SL Don Hearth had already appointed the experienced New-ton W. "Bill" Cunningham as MVM program manager and the ener-getic geologist Steve Dwornik as program scientist. JPL was comfort-able with Cunningham from his days as Ranger program manager and knew he would represent the project well at NASA Headquarters.

In January 1970 the 1973 Mariner 10 Venus-Mercury project office was officially established at JPL with Walker E. "Gene" Giberson as the project manager. Bill Pickering's and Bob Parks's selection of Giber-son was all-important to the management of this project. Most of the spacecraft work was going to be done by a systems contractor, which JPL did not relish, and the budget was an almost impossibly tight $98 million, guaranteed "in blood" by the big boss, Dr. Pickering. Into this losing situation charged dynamic Gene Giberson. Gene had been kicked out of the project manager's chair when JPL's only other proj-ect with a systems contractor had run into trouble (Hughes on the Sur-veyor project), and he was determined to restore his reputation. If he could bring in a successful mission within that impossible budget it would establish him as an outstanding project manager. I believe mo-tivation is the most important aspect of management, and Gene was fully motivated.

So was the contractor. After a lively competition the Boeing Com-pany was selected as the systems contractor. Boeing believed as I did that we could surely sell a gravity-assisted Grand Tour of the outer planets in the late 1970s. They wanted to be the systems contractor on that ambitious project. If they won the contract for Mariner 10, they would work on the first gravity-assist mission and would establish their team with the people at JPL, so Boeing management organized a first-class effort.

We had made it very clear in a preproposal briefing to the competi-tors that the budget for Mariner 10 was going to be extremely tight. Northrup Corporation took this too much to heart and proposed to design the spacecraft with mostly low-cost draftsmen and only a few engineers. This was immediately rejected by JPL, and I concurred.

Boeing showed convincing data that, during a recent slowdown in sales and deep layoffs at their Seattle-area plants they had not only kept overhead from increasing, as was normal during slowdowns, but actually cut it in half. Moreover, they presented convincing statistics that there had been no loss in operating efficiency. Very impressive. We figured they could operate within the budget if anyone could, and a contract was signed in July 1971. Nevertheless, Boeing still had to convince the JPLers that they could design and build a spacecraft as well as or better than JPL.

It was fun to watch this process. Boeing had an excellent management pair in blunt and forceful Ed Czarnecki as project manager and sharp but more diplomatic Haim Kennet as his deputy. Giberson had assembled a small but very capable team at JPL, including John Casani, Jim Wilson, Norri Sirri, Vic Clarke, Jim Dunne, Clayne Yeates, Nick Renzetti, Ek Davis, Gael Squibb, and Dallas Beauchamp, all of whom would go on to play leading roles in future JPL projects. None of these people were afraid to speak their opinions, which reflected Giberson—you could always count on Gene to give it to you straight, just what he was thinking.

Meetings between the JPLers and the Boeing engineers were almost literally a riot. They would argue fiercely all day over a spacecraft design feature. Finally Boeing would get JPL to say that it had to be done their way because "that's how we always do it." "Aha," the Boeing people would snap back, "there is no good technical reason to do it that particular way, you are just used to it. Our way will work better and cost less." The teams would eventually reach an agreement and then go out for beer and dinner like a bunch of old-time buddies. The next morning they would be on to another spacecraft feature and the shouting would start all over again.

All in all this was a very effective working relationship. Although JPL had the reputation of being tough and dictatorial with its contractors, under the leadership of Giberson and Czarnecki, JPL and Boeing stood toe-to-toe and debated until they jointly came to the best solution. During my occasional visits to Boeing to check on progress I always came away with a good feeling. There were no drawn-out delays in making decisions. In that regard Gene Giberson was like all successful project managers: they have great confidence in their own judgment, given the proper facts, and actually relish making tough decisions.

For further insight into what kind of people make good project managers for pioneering ventures into space, perhaps one side activity on this project will give a clue. A group of the JPLers, led by Bob Parks and John Casani, decided that they should climb Mount Rainier while they were in the Seattle area working with Boeing. Rainier was not something to be taken lightly. Sixty-seven climbers had died in the attempt to scale it, most often due to sudden unexpected summer storms on the mountain. Sure enough, the JPLers got caught in a nonseasonal blizzard—visibility was close to zero, and they were extremely fortunate to get back down the mountain. Give up? No way. The next year they were back again and made a successful climb to the summit. Determination is part of the project manager's character.

Science on a Shoestring

Gene Giberson had a fully motivated contractor in Boeing, but he was going to need more than that to have any chance of staying within Pickering's $98 million ceiling. First of all he decreed that existing subsystem designs and hardware had to be used wherever possible. Most subsystem designs came from prior Mariners, but the hydrazine propellant tank design for the midcourse propulsion subsystem was borrowed from Pioneer 10/11 because it had a greater capacity. After allocating minimum budgets to all subsystems and integration/test functions Gene found he had only 3 percent reserve rather than the 10 to 15 percent considered the minimum for even a conservative project. And Venus-Mercury was not exactly conservative.

For one thing, in going more than half way toward the Sun the Mariner 10 spacecraft would experience almost five times the solar intensity as at Earth. Once deployed after launch, the two solar panels had to be gradually tilted around their longitudinal axis to prevent overheating from the increasing solar intensity. To protect the body of the spacecraft the design team was studying combinations of reflectors, louvers, and thermal blankets. In his excellent book *Far Travelers: The*

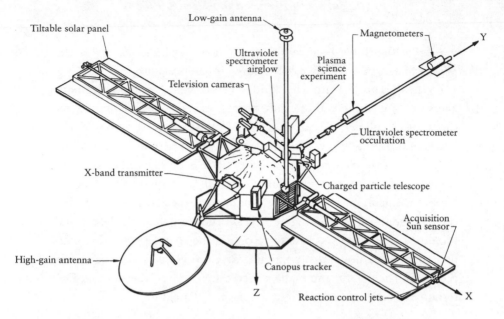

Low-gain antenna

Tiltable solar panel

Magnetometers

Y

Ultraviolet
spectrometer
airglow

Plasma
science
experiment

Television cameras

Ultraviolet spectrometer
occultation

X-band transmitter

Charged particle telescope

Acquisition
Sun sensor

High-gain antenna

Canopus tracker

Z

Reaction control jets

X

On 3 November 1973, the Mariner 10 spacecraft was launched on the world's first gravity-assist mission, taking the first closeup images of Venus and making three passes of Mercury. (Courtesy NASA)

Exploring Machines, Oran Nicks gives me credit for another solution: "One of the solar protective devices was an umbrella-like sunshade made of a Teflon-coated glass fiber fabric known as beta cloth. This simple device, suggested by Robert Kraemer of NASA Headquarters, unfurled in the same manner as an umbrella, and shadow shielded the rocket system and parts of the spacecraft when pointed toward the Sun." I remember suggesting an umbrella, but I am not at all sure I was the first to think of it. In any event, it worked.

There was the usual fierce competition among planetary scientists to be selected as members of the Mariner Venus-Mercury science team. Code SL tried to keep the selection as fair as possible and based on technical merit rather than political clout. To be selected, the Space Science Steering Committee, comprised of the science chiefs throughout OSSA (Headquarters Code S) and chaired by the science deputy for Code S, had to give a proposed instrument and its team a rating of Category 1, meaning it had strong scientific value and was deemed developmentally ready by the Project Office at JPL. In the imaging competition there appeared to be an unfortunate bias. The imaging instrument itself was set, because on the tight budget we could only afford to use one or two of the already-developed vidicons employed on Mariner 9, so the competition was for their best use and methods of image processing and interpretation. As usual, the proposal from Bob Leighton, who had led the imaging team on Mariners 4, 6, and 7, was rated as Category 1. Code SL scientists were pleased to see that Leighton would get some real competition from a strong proposal from Bruce Murray of Caltech teamed with Merton Davies, a space reconnaissance imaging expert from the Rand Corporation. However, Urner Lidell, a member of the SSSC, thought imaging was low-grade science and was especially down on Bruce. He wanted the SSSC to rate Murray's proposal as Category 3, that is, second-rate science. Don Rea and Steve Dwornik, Code SL's irrepressible science chief for Planetary Geology and the MVM program scientist, alerted the other members of the SSSC, including Bob Fellows (Planetary Atmospheres), Dick Young (Exo-Biology), and Bill Brunk (Planetary Astronomy), and managed to get Murray's proposal a fair hearing. Murray's proposal was finally judged by the full SSSC to be Category 1; it made the final selection.

Now, Bruce Murray, as well as being young and vigorous, was then and still is not known for timidity. After selection he promptly showed up at NASA Headquarters and pointed out the great difficulties his team was facing in trying to live with the Mariner 9 vidicon imaging system. Mariner 10, picking up speed as it fell toward the Sun, would fly by Mercury at an unprecedented fifteen thousand miles per hour. There would be time to snap only about thirty shots of Mercury at poor resolution and load them into the spacecraft tape recorder for later slow playback and transmission. To get around the tape-recorder limitation Bruce stated that his team would need to develop a film system with image motion compensation, similar to the Lunar Orbiter system. Price, astronomical—about $57 million, as I recall. I pointed out that our total budget for Mariner 10 science was only $10.8 million, from which imaging was allocated a lion's share $6 million, and that had to include not only the camera but also all the processing, photo prints, and at least the initial analysis.

Bruce was undaunted. He came back shortly thereafter proposing a new dielectric camera being developed by RCA. The cost was a bit uncertain but estimated to be about $40 million. I had to tell Bruce again that his budget was $6 million, firm, no appeal. It was clear that the imaging team would have to stick with the existing imaging system from Mariner 9, which flew two cameras, one with a wide-angle lens and one with a 500-millimeter telescopic lens. But that still meant only thirty snapshots.

Fortunately Bruce had worked with JPL and knew and respected the Mariner 10 systems engineers. They put their heads together and came up with an ingenious solution. If only one-quarter of the pixels (picture elements) were returned from each frame, the data could be transmitted to Earth in real time at the bit rate allotted to imaging, which they had managed to get boosted to 118,000 bits per second by talking the JPL communications engineers out of some of their conservative margins. With two vidicon cameras with upgraded 1500-millimeter lenses taking one-quarter-sized pictures, one camera could be taking a new picture while the other was being read out and transmitted. More than three thousand images could be returned during the fast flyby, and a mosaic of the entire lighted surface could be assembled from the postage-stamp high-resolution images. An elegant solution and within the $6 million budget.

The only thing missing was the desired wide-angle shot of the entire planet. However, the team, now including JPL's Ed Danielson, came up with another ingenious solution. Each camera from Mariner 9 had an eight-position filter wheel to allow imaging in various wavelengths. They simply substituted a mirror in place of one of the filters. The mirror could deflect light up to a very small and simple top-mounted wide-angle lens and, voilà, one had the wide-angle image at no appreciable extra cost. Very clever design. It never ceases to amaze me what motivated people can accomplish.

There was another fierce competition between the magnetometer teams of Ed Smith at JPL and Norm Ness at Goddard. Both teams had excellent, well-developed magnetometers and were technically very strong. Norm was a real bear to deal with, whereas Ed was mister nice guy, but we were going to stick to our principle of selecting on technical merit, not personalities or politics. Ed's team had won in our most

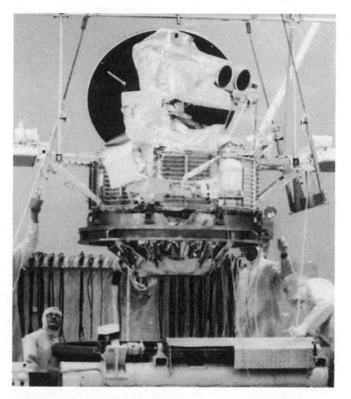

Twin vidicon cameras with folded-optic Cassegrain telescopic lenses provided high-resolution imaging on Mariner 10. Mounted on top of each camera is a small lens for wide-angle imaging. (Courtesy NASA)

recent selection for Pioneers 10 and 11, and it was on a roll. Norm, though, had come up with a simple but clever idea. One difficulty with measuring low-level magnetic fields was to filter out the flux generated by electric currents within the spacecraft. Putting the magnetometer out on a long boom lessened the interference but did not eliminate the problem. Norm proposed mounting a second magnetometer half way out on the boom. Comparing the differences in the readings of the two magnetometers would permit isolating the effects of the spacecraft. Simple but ingenious. It was a winner. (It later won again in the competition for Voyager, much to Ed Smith's frustration.)

The final outcome of the science competition was as follows:

Experiment	*Team Leader and Principal Investigator*
Television	Bruce Murray
Plasma science	Herbert Bridge
Ultraviolet spectroscopy	Lyle Broadfoot
Infrared radiometry	Stillman Chase
Charged particles	John Simpson
Radio science	H. T. Howard
Magnetic fields	Norman Ness

Knowing that his biggest battle on MVM was to live within that skimpy budget, Giberson made one especially bold move to contain costs that made me very nervous. He decreed that there would be no detailed design or hardware procurement until eighteen months before launch. That is certainly one way to control costs, as the team could charge only a limited number of man-hours in just eighteen months, but it meant that even just six months before the scheduled launch we would still not have an assembled spacecraft. Kind of spooky.

Once into the hardware-development phase, there was a good deal of testing going on at Boeing and JPL, but the work at Boeing had a different look. At both JPL and NASA it was customary for a substantial group of engineers, technicians, and inspectors to attend each major test, whereas at Boeing the only people in attendance were the design engineer, the test engineer, maybe one technician, and a guy with a clipboard who was the official inspector as well as data recorder. In answer to JPL concerns Czarnecki just asked, "Why do we need more?" He won that one too.

Getting a Quality Atlas

Our Mariner 10 spacecraft had come together nicely, but I was becoming more concerned about the Atlas-Centaur launch vehicle. All of our prior Mariner mission failures—Mariners 1, 2 (partial failure), 3, and 8—had been due to faults in the Atlas launch vehicles. Fortunately, for each of those failed launches we had a backup spacecraft and launch vehicle that succeeded, but for low-budget Mariner Venus-

Mercury we had only one precious spacecraft—we could not afford a launch failure.

George Newton, the Atlas program manager at Convair, was bluntly pessimistic. He said their assembly-line employees had been told by Convair management that this was the last Atlas-Centaur currently on order and that they would probably all be laid off when it was completed. Newton said morale was terrible, and he recommended that NASA tear down and reassemble the entire vehicle after it reached Cape Canaveral. Great. When I suggested that our project people come to their plant in San Diego and brief the Convair employees on the importance and exciting nature of the MVM mission, Newton replied, "Forget it, these people aren't going to respond to any pep talk." I insisted we try it nonetheless. Jim Dunne, our MVM project scientist at JPL, took a group of our mission scientists down to San Diego and they described what they were doing and what they hoped to learn. Their enthusiasm proved to be infectious. Afterward George Newton expressed his amazement; he said the assembly people were intrigued and motivated. This vehicle was delivered on schedule and checked out well at the cape.

A Jinxed Launch?

Meanwhile the flight spacecraft passed all its formal and rigorous reviews by JPL and NASA Headquarters and was shipped on schedule to Cape Canaveral for launch preparations. At this point I proposed to George Low, NASA's deputy administrator, that he give the MVM team a special award. George had been campaigning for lower project costs, but project managers all knew that nothing succeeded like success. If a mission were successful, any cost overrun was quickly forgotten, but if a mission failed, it mattered not one bit if it was launched within its budget. I advised George that if he wanted to change this pattern and try to motivate project managers to control costs then here was an excellent opportunity. Here was his chance to reward a project team that had delivered a quality spacecraft on time within an extremely tight budget and *before* the mission either succeeded or failed. George responded, "No way." He was not going to have a mission failure blamed on Headquarters imposing a too-tight budget. He felt that would undermine his campaign to reduce project costs.

A few days later I ran into George as he was waiting to catch a flight at Washington's Dulles Airport. I made one more appeal. To my surprise (George had a will of iron and rarely reversed his decisions) he backed down and gave his OK, but he warned me that if Mariner 10 failed for any reason he was personally going to lead a review team that would prove the fault was some error by the project team and not because of a tight budget. George always played hardball. Anyway, before launch the Mariner 10 team was indeed given the Presidential Award for Management Excellence, the first NASA had ever received. I was proud of them, but Gene Giberson was furious with me—he was sure the award before launch would be a jinx. I think we were all superstitious about launches, so I did not blame him for feeling that way. Nevertheless, I felt the team should be rewarded for a fantastic job of management.

I was not going to contribute any more to jinx the launch, so I went through my usual ritual of having raw oysters at Captain Ed's Seafood Restaurant at the cape the evening before the launch. (What, me superstitious?) Because our spacecraft was to head toward the Sun, the launch would be in the early morning, soon after midnight, to take advantage of Earth's rotation in canceling some of Earth's velocity around the Sun. Night launches are a beautiful sight. The Atlas-Centaur out on its pad looked like a brilliant white-and-silver lighthouse standing out against the jet black background over the deserted Atlantic Ocean. Liftoff occurred right on schedule, at 12:45 A.M. on 2 November 1973. The Atlas engines rattled the windows and lit up the sky with their fiery blowtorch exhausts. Mariner 10 was on its way. Then the fun and games began.

The Perils Begin

As the science instruments were turned on after launch, the heaters for the television cameras would not come on. With the all-too-effective beta-cloth sunshade blocking out sunlight the cameras were going to get very cold. The fear was that thermal contraction would crack the

potting compound in the vidicons and result in total failure of the imaging subsystem. Effective insulation prevented heat from the body of
the spacecraft from reaching the cameras on their isolated scan platform. All that could be done was to leave the vidicons turned on for
the entire trip to Venus so that they might benefit from internally generated heat. But then, just two weeks before Venus encounter, with the
sort of quirky behavior that would characterize the entire mission, the
camera heaters miraculously came on of their own accord.

The next major anomaly was an uncommanded and irreversible
switch to the backup power mode. If whatever fault caused this switch
occurred in the backup subsystem, the mission would come to an
abrupt end. Great care was taken to baby the power supply for the re-
mainder of the mission.

Next the feed on the high-gain antenna failed, a situation that, un-
corrected, would doom the real-time imaging capability that had so
cleverly been engineered into the mission. Four days later it came back
on all by itself. Ah, but four hours later it was off again. Analysis gave
hope that higher temperatures as the spacecraft approached the Sun
would cure this open circuit, and that indeed proved correct.

Just a week before Venus encounter came a real killer. The spacecraft
started into an unexpected oscillation about its roll axis. The gaseous
nitrogen attitude-control jets tried valiantly to counter the rolling mo-
tions, but in doing so were dumping gas at a disastrously high rate.
Back at JPL a tall, mustachioed young engineer by the name of Bill
Purdy, who was the Guidance and Control analyst, suddenly became
the focal point. He and his team quickly decided to shut down the gyroscopes that were commanding the nitrogen jets to come on, but in
the short hour it took to accomplish that shutdown one-sixth of the entire nitrogen gas supply had been depleted. Every bit of the rest was
going to be needed to complete the full mission.

Venus Unveiled

A decision was made to control the attitude of the spacecraft during
the Venus flyby by locking the spacecraft onto the Sun and the star

Canopus, rather than by controlling attitude with the gyros as originally planned. There was the risk that the Canopus sensor would be
fooled by reflected light from Venus, but that did not happen. In fact,
the science mission at Venus went beautifully. More than four thousand
images were taken of the Venusian cloud tops, the images in the ultraviolet revealing for the first time intricate swirls and a fantastically high
cloud velocity that sent clouds circling completely around the planet in
just four days. Prof. Verner Suomi, a senior member of the Imaging
Team and one of the deans of meteorology worldwide, was enormously
excited by these pictures. With his Scandinavian blue eyes gleaming, he
predicted that Mariner 10's full image of Venus would be used as the
frontispiece for every book on meteorology published in the next one
hundred years.

Vern was a bit carried away in his enthusiasm, but his excitement
stemmed from finally seeing a Hadley cell. In 1735 an English gentleman named Hadley had published a theory on what caused the all-
important trade winds that allowed ships to sail westward before the
wind across the Atlantic in low latitudes in the tropics and then return
before easterly winds at higher latitudes. His theory was based on rising warm-air currents near the equator traversing at high altitude to
the poles, cooling and falling to the surface, then migrating back to the
tropics. A problem was that Earth is so complicated by its mixture of
water and land masses, rapid rotation, and mountain ranges, that even
today's clear satellite images of the entire Earth show a complex hodge-
podge of swirling clouds rather than the expected large cell patterns
as predicted by Hadley. But on less complicated Venus, with no oceans
and a very slow rotation rate (243 Earth days for one rotation), there
in the UV images as clear as Hadley's sketches was the long-predicted
Hadley-cell pattern in the clouds. Vern Suomi was practically doing
handsprings.

The excitement at Venus encounter was something to relish, but we
still had to get to Mercury without using up the attitude-control gas.
The cause of the roll oscillation was pinned down to unexpected oscillation of the long magnetometer boom in spite of design steps taken
to prevent this possibility. The magnetometer PI, steely-eyed Norm
Ness, had bargained hard for the longest boom he could get to place

A Mariner 10 ultraviolet image of Venus, revealing a classic Hadley cell circulation pattern in the Venusian clouds, confirming a meteorological theory first advanced in 1735 to explain Earth's trade winds. (Courtesy NASA)

his sensors as far as possible from the magnetic perturbations of the spacecraft. For Mariner 10 he got the boom length up to six meters (twenty feet). The boom dynamics were tested on the ground, but it is impossible to exactly simulate zero gravity, so Norm's boom had to be supported during testing, somewhat compromising the validity of the test. In space the boom's natural frequency got in sync with the gyro system and the nitrogen jets started fighting the oscillations on every roll.

A wild proposed solution was to try solar sailing, using the photonic pressure of sunlight and the ionized solar wind to control the spacecraft attitude. This had been tried on Mariner 4, with moveable paddles on the ends of the four solar panels, but it was not effective due to the reduced solar pressure as the spacecraft progressed outward to the orbit of Mars. Going toward the Sun from Venus to Mercury, the increased solar pressure just might be enough to orient the spacecraft. It was! JPL controllers were able to vary the tilt of the spacecraft's solar panels and the fully pointable high-gain antenna to act as fins and a rudder to stabilize the spacecraft without using the precious nitrogen gas, saving it for critical maneuvers in which both the gas jets and the gyros had to be used.

Mariner 10 presented more than its share of problems to solve, but it made it to Mercury and executed its science mission there exactly as planned.

Mercury Close Up

On 29 March 1974 Mariner 10 whizzed by Mercury with its two cameras snapping pictures like a two-barreled Gatling gun, a new image every forty-two seconds, more than two thousand in all. Back at JPL in the imaging processing lab the technicians and Imaging Team members worked at a furious pace to assemble the miniframes into large photo mosaics. The first ever closeup view of Mercury was fascinating—much like the Moon, with basins, highlands, and plains, all liberally peppered with craters of varying size but also with some distinctly nonlunar features. For one thing, there were large scarps or cliffs nearly three kilometers (two miles) high and stretching as far as five hundred kilometers (three hundred miles). These thrust faults appeared to be the result of compression in the surface as the forming planet cooled. A count of the small craters on Mercury revealed about the

A photomosaic of Mercury, constructed of eighteen photographs taken by Mariner 10 during its first of three flybys of Mercury on 29 March 1974. The edge of a large circular basin named Caloris, about thirteen hundred kilometers (eight hundred miles) in diameter, is seen along the day-night terminator at the left center of the mosaic. (Courtesy NASA)

same density as on Mars and the Moon, indicating that meteorites were spread evenly over the inner solar system, at least during the later stages of planet formation.

One enormous impact basin, thirteen hundred kilometers (eight hundred miles) across, was right at Mercury's equator and immediately caught one's eye. Geologists compared it directly to the somewhat smaller Mare Imbrium basin on the Moon. This giant basin, Mercury's most prominent feature, was named Caloris (the Greek word for "hot") because it directly faces the Sun at every other perihelion (closest approach).

Mariner 10 found that Mercury possessed a tenuous atmosphere of helium, but so thin that day and night temperatures were extreme, plunging from 460 degrees Kelvin (369 degrees Fahrenheit) at midday to 90 degrees Kelvin (−297 degrees Fahrenheit) at night. The most unexpected finding was the detection of a magnetic field, only one-sixtieth the magnitude of Earth's but definitely there. The prevailing theory was that a planet had to rotate much faster than did Mercury to generate an internal magnetic field. Was Mercury's field just due to surface electrical currents induced by the solar wind? One flyby was insufficient to provide a conclusive answer.

It should be noted that the mission scientists were under a good deal of pressure at the time of planet encounter, some of it imposed by NASA Headquarters. In the early days of space science missions the PIs were given up to a year to analyze their data and release results. That changed, however, as soon as cameras were place aboard the spacecraft and Code SL responded to urging from the news media personnel for the quick release of new images from the spacecraft cameras. After all, that was what the public found to be most exciting and what helped generate enthusiasm and support for planetary exploration. So SL made all its PIs generate computer programs for quick analysis of their data so that they could present results days or even hours after a planetary encounter. Occasionally this led to shooting a little too quickly from the hip.

As an example, as the Mariner 10 spacecraft approached Mercury, Lyle Broadfoot's UV spectroscopy team discovered a bright spot in the background. Broadfoot, who always impressed me as being cautious, nevertheless wanted, as did all PIs, to make a splash at the daily press conferences. He convinced his other team members, including the young but very sharp and highly respected Michael McElroy, that what they were seeing was a satellite of Mercury, and that was the announcement he made at the next press conference. The JPL astro-navigators got right on the data and were quickly able to identify the shining object as just a star in the background, not a Mercury satellite. So "instant science" produced a temporary erroneous conclusion, but the merits of releasing results quickly still far outweighed false starts.

Summarizing the mission up to this point, Oran Nicks said, "Like an unruly child who behaves very badly and becomes a model child just as anxious parents expect to be embarrassed, Mariner 10 began to function perfectly again just prior to its encounter with Mercury. . . . High-resolution photographic coverage of Mercury was achieved as planned. . . . Shortly after encounter, in now typical Mariner 10 fashion, problems began to recur."[2]

It was important to keep the spacecraft alive because we had the opportunity for more encounters of Mercury. Back in 1970 Prof. Giuseppe Columbo of the Institute of Applied Mechanics in Padua, Italy, had pointed out that after Mercury encounter the spacecraft orbital period about the Sun could be made to be 176 days, exactly twice Mercury's orbital period of 88 days, so revisits to Mercury would be possible. And that was our aim.

Mercury Again and Again

For a detailed look at all the problems, and solutions, between encounters, the reader is referred to the list of references, especially *The Voyage of Mariner 10*, by James Dunne and Eric Burgess. Mariner 10 continued its pattern of crises between planets then nothing but beautiful science during encounters. It made three visits to Mercury, including a very close third pass at only 620 miles altitude. One week later it ran out of its precious nitrogen gas and was turned off forever. That third and final close pass was exceptionally important, for it confirmed that Mercury did indeed have a magnetic field of internal origin. Norm Ness's troublesome magnetometer boom had produced one of the greater discoveries of the MVM mission. So much for the supposed requirement for rapid planet rotation to develop a magnetic field. Moreover, current theory says that such a magnetic field comes from a large, molten, thermally convecting core, and yet the sharp photo mosaics of the heavily cratered surface produced from the Mariner 10 cameras show no evidence of associated volcanism on Mercury in the past three billion years. Another puzzle to challenge the theorists and cosmologists.

One frustrating limitation to the imaging was that we were able to see only one hemisphere of Mercury. Mercury has a so-called dumbbell rotation, in that it rotates one and one-half times in each revolution about the Sun so that at each perihelion it presents an opposite face. Mariner 10's orbit took twice as long as Mercury's orbit, and as a result, each time the spacecraft reached Mercury the planet had the same hemisphere lit by the Sun. No matter how many times we could nurse Mariner 10 back to Mercury, we were going to see the same lighted face of the planet. It is not surprising that there are currently serious proposals to send another (mini) spacecraft back to study Mercury long-term from an orbit around the planet.

For the record books, Mariner 10 was the first spacecraft to take closeup images of Venus, the first to do a gravity-assist mission, and the first to explore the planet Mercury. A record of which to be proud.

Were all the in-flight problems due to sloppy engineering or inadequate testing because of a too-tight budget? I saw no conclusive evidence of that. Testing was very thorough, albeit without frills. It was just a challenging mission, facing many new design problems, and it took lots of human ingenuity and dedicated effort to make the mechanisms surmount unexpected conditions. I am very fond of this project, even though I called the spacecraft "Pauline," in reference to the silent movie serial in which each episode ended with the heroine Pauline in a hopeless situation and each new episode began with her miraculous escape. To me the Mariner 10 mission will always be *The Perils of Pauline.*

1975

VIKING I AND 2 MARS ORBITS AND LANDINGS

For centuries humans have dreamed of the possibility of life on another planet. Perhaps it would be intelligent life, advanced beyond the current state of the human race. Certainly those dreams intensified in 1877, when Italian astronomer Giovanni Schiaparelli announced that he had observed a network of linear features on the surface of Mars. The Italian word he used for these features was *canali,* meaning "grooves" or "channels," but the English-speaking news media translated it as "canals," leading to worldwide speculation about intelligent creatures on Mars.

In 1894 a well-to-do Bostonian, Percival Lowell, educated as a mathematician and fascinated by astronomy, went so far as to construct and staff a first-class observatory in Flagstaff, Arizona, dedicated to observations of Mars. As had Schiaparelli, during periods of exceptional viewing he also reported seeing linear features and intriguing color changes, including some areas that changed seasonally to a dark blue-green. As telescope observations improved, Mars was shaping up as a "twin" of Earth's in many fascinating ways. For example, Mars not only has an Earth-like twenty-four-hour day but also has its axis of rotation tilted twenty-five degrees, similarly to Earth's, producing seasonal changes during the Martian year. Lowell believed that the seasonal darkening he saw was vegetation turning green during the Martian spring.

NASA scientists Christopher McKay and Wanda Davis have nicely summarized the turn-of-the-century assessment of Mars:

> Early telescopic observations revealed Earth-like seasonal patterns on Mars. Large white polar caps that grew in winter and shrunk in the summer were clearly visible. Regions of the planet's surface appeared to darken beginning near the polar cap that was at the start of its spring season and spreading toward the equator. It was natural that these changes, familiar to patterns on Earth, would be attributed to similar causes. Hence, the polar caps were thought to be water ice and the wave of darkening was believed to be caused by the growth of vegetation. The 19th century arguments for the existence of life, and even intelligent life, on Mars culminated in the book *Mars as the Abode of Life* by Percival Lowell in 1908.[1]

Lowell had to admit that he observed no oceans or lakes on Mars and that there was a "dearth of water" on the planet, but to him that explained the need for canals to carry precious water from the melting polar caps. Admitting that to be seen from Earth the canals would have to be inordinately wide, Lowell explained that "what we see . . . is not really the canal at all, but the strip of fertilized land bordering it."[2] The canal itself might be too narrow to be seen from Earth, but the canal plus the dark vegetation areas on each side of the canal would be observable. The press was delighted to print Lowell's conjectures about a civilization of highly intelligent Martians actively irrigating and growing crops on Mars. Writers such as Edgar Rice Burroughs, H. G. Wells, and, later, Ray Bradbury further fanned the public interest with imaginative tales of alien Martians attacking humans.

In the late 1950s observations using new infrared spectroscopy techniques with the Palomar two-hundred-inch telescope showed spectral lines of chlorophyll from Mars, which added credence to the presence of vegetation on the Martian surface. These readings were later found to be coming from water in Earth's atmosphere, but they caused excitement at the time. By 1960 Mars was known to have a very thin, oxygen-free atmosphere, and I do not believe there were any accredited scientists who still maintained that the planet was home to a civilization of intelligent beings. Nevertheless, there was moisture on Mars, and microorganisms and even plant life could have evolved there. It was agreed that Mars was still the most likely place in our solar system to search for extraterrestrial life. Fred Whipple, then chairman of the Department of Astronomy at Harvard University, summed up the scientific assessment:

> Chances are that bacteria are the only type of animal life which could exist in a planet's oxygenless atmosphere. There also may be some sort of tough primitive plant life—perhaps lichens or mosses which produce their own oxygen and water. Such plants might explain the changing colors of the Martian seasons.
>
> There's one other possibility. How can we say with absolute certainty that there isn't a different form of life existing on Mars—a kind of life that we know nothing about? We can't. There's only one way to find out for sure what is on Mars—and that's to go there.[3]

The technology was available in the 1960s to do just that—send robot spacecraft to explore Mars.

Beginning in 1962 the Soviet Union launched multiple Mars spacecraft at every twenty-six-month Mars opportunity, but without success. The United States had better luck, and its pioneering 1964 Mariner 4 spacecraft successfully made it to a Mars flyby in July 1965. Its television camera took twenty-one small images of the surface at a respectable resolution. The images were stored in an on-board tape recorder and then transmitted to Earth at the agonizingly slow rate of $8\frac{1}{3}$ bits per second, taking about twenty-five minutes per image. As those images appeared at JPL I think that most of us, at least the non-astronomers, expected to see some explanation for Schiaparelli's and Lowell's linear features—not water-filled canals, but at least rift valleys or dark ridges. Nothing. Only a heavily cratered ancient surface, much like the Moon. Really quite a letdown.

The NASA-JPL team followed up with Mariners 6 and 7, launched in 1969, with much more advanced cameras and instruments and a data-transmission rate of up to 16,200 bits per second. Its images of the Martian south polar cap were fascinating but again showed an ancient Moon-like cratered surface with no green areas and no canal-like features. Moreover, the spacecrafts' infrared instruments determined that the south polar cap was frozen carbon dioxide, not water ice. As Caltech scientist Bruce Murray said, "Many who had been hopeful of life were pretty discouraged, because a frozen, dry moon with a carbon dioxide envelope seemed to be a very unpromising abode for life."[4]

Our knowledge of Mars could have stagnated here for who knows how long. Fortunately, the American people and their representatives in Congress supported our Mariner 9 spacecraft, which in 1971 mapped the entire surface of Mars from orbit. The resulting image mosaics revealed that the southern hemisphere of Mars, observed by Mariners 4, 6, and 7, was indeed mostly ancient and heavily cratered, but in contrast the equatorial region and the northern hemisphere had gone through relatively recent and wild geological activity, including massive flooding and erosion. Mars was far more exciting than the Moon and could very well have developed life-forms during its wet period.

Well before the results from Mariner 9, a priority long-range goal of all involved in planetary exploration was to land on Mars, and not just for public appeal and the thrill of a big first but in accordance with one of NAS's three goals of planetary exploration: to better understand the origin and evolution of life. By the mid-1960s scientists knew that the evolution of complex life-forms on Mars was highly unlikely, due to low atmospheric pressure and temperature ruling out liquid water on the surface, but the presence of simpler microorganisms was a possibility. The only way to find such organisms was to get right down on the surface. It was up to NASA Headquarters to sell a Mars lander project.

The Short Life of Voyager Mars

Since its inception in 1961 NASA's Office of Space Science[5] had been headed by Dr. Homer Newell. Bespectacled and crew-cut Homer came close to my picture of an ideal scientist, exhibiting intelligent understanding of complex subjects and reflecting calm and dignity even in stormy times. He was fortunate to have as his deputy Edgar Cortright, one of the most talented and articulate of all of NASA's engineering managers. Homer preferred to work with his fellow scientists, not especially with engineers and certainly not with contractors and marketeers, but Ed could and would work effectively and constructively with all participants, including congressmen and business representatives. Together Homer and Ed made a formidable team. In the light of strong support for the Apollo Program and for winning the space race, stoked by persistent Russian attempts to land on Mars, they believed they could and should sell a very ambitious program to land and search for life on Mars and then proceed to land on and explore the surface of Venus. They called this proposed program Voyager (a name that would reappear a decade later).

Voyager planning got underway with a number of funded studies involving JPL, Langley Research Center, Ames Research Center, and their contractors. A Voyager program office at Headquarters was started with Oran Nicks as director (in addition to his continuing role as director of Lunar and Planetary Exploration) and Don Hearth as deputy director. In the spring of 1967 Oran recruited me as program manager for the Voyager Mars Surface Laboratory. As chief engineer for Space Systems and program manager for Lunar and Planetary Programs at Ford Aeronutronic I had been overseeing a NASA-funded design effort on an automated biological laboratory (ABL) as a compact but comprehensive payload to search for life on the Martian surface, so I was very much up to speed on the efforts of NASA and its contractors to sell the Voyager program.

I accepted Oran's offer even though I had strong doubts that even the talented duo of Newell and Cortright could get the program funded. There were several major problems. For one, the budget for NASA was under severe strains. Apollo costs had grown, magnified by the tragic Apollo 1 fire, the United States was ensnared in a very costly and unpopular war in Vietnam, and domestic unrest was popping out in riots across the land. To make matters worse, Voyager's size had grown completely out of hand. The Voyager orbiter-lander-rover payload had been scaled to be launched on a Saturn 1, but in October 1965 NASA decided that there weren't enough missions to justify the continued production of the Saturn 1, so that left the Saturn 5 as the only launch vehicle with adequate capability. The Saturn 5, though, was just too darn big. To use the enormous payload capability of the Saturn 5, which could propel over sixty thousand pounds to Mars, the Voyager designers gave everyone a generous weight allowance, yet it was still possible to carry two complete orbiter-lander-rover spacecraft assemblies on each launch. Even in 1968 dollars Voyager Mars would surely have cost at least $5 billion. The program was delayed and then finally rejected for good by Congress in October 1967. American taxpayers were well represented.

Recovery from Voyager Mars

The fall of 1967 saw new assignments at NASA Headquarters. Homer Newell was promoted to associate administrator, John Naugle replaced Homer as head of OSSA with Oran Nicks as his deputy,[6] and Ed Cortright moved to OMSF as deputy associate administrator. Don Hearth became the new director of Lunar and Planetary Programs, and I was appointed his manager of Advanced Planetary Programs and Technology, charged with planning future missions for our program.

But what program? NASA had put all of its planetary eggs in the one large Voyager basket. Moreover, Jim Webb, NASA's administrator, was angry that Congress had rejected his proposed Voyager project and indicated that he was going to emphasize the loss by not proposing any new planetary projects for at least two years. He made more expansive noises to the congressional committees, noting that we still had two Mariner spacecraft in assembly for launch to Mars flybys in 1969, but all he would let us substitute for Voyager in the fiscal year 1968 appropriations and the fiscal year 1969 budget request then in preparation was a meager $6 million per year in an effort to "keep JPL alive."

Don Hearth and I quickly decided that we would not take Webb's directions too literally and would use every penny of that $6 million to create new planetary projects. Now that sounds like deliberately disobeying a direct order from the boss, so it needs a bit of explanation. I think everyone in NASA had great respect for Jim Webb, not just for his ability to make big and bold decisions but for his intimate knowl-

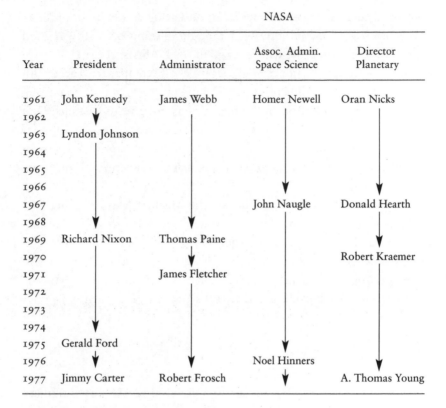

Year	President	NASA Administrator	Assoc. Admin. Space Science	Director Planetary
1961	John Kennedy	James Webb	Homer Newell	Oran Nicks
1962				
1963	Lyndon Johnson			
1964				
1965				
1966				
1967			John Naugle	Donald Hearth
1968				
1969	Richard Nixon	Thomas Paine		
1970				Robert Kraemer
1971		James Fletcher		
1972				
1973				
1974				
1975	Gerald Ford			
1976			Noel Hinners	
1977	Jimmy Carter	Robert Frosch		A. Thomas Young

U.S. presidents and key NASA managers from 1961 through the golden era of the 1970s.

edge of the budget process within the White House and on Capitol Hill, giving him a unique ability to sell new space projects. As his deputy, Robert Seamans Jr., would later recall, "Jim was the charismatic leader with long-range vision and a great knack for understanding how policy and politics interacted in Washington."[7] The problem was in communications. Webb lacked the technical education to understand all the technical jargon his scientists and engineers used in their presentations to him. They in turn did not comprehend much of what he was trying to achieve. Reference the computer-buy story back in chapter two in which the engineers naturally thought they were supposed to procure the best large mainframe computer for their mission but Webb was determined to award the contract to an underdog company in order to give IBM some competition. Webb thought in terms of what was good for the country, not just what was good for NASA and the U.S. space program—something not clear at that time to his managers. Faced with sometimes puzzling directions, NASA's managers were used to putting their own interpretation on Webb's directions. In this instance Don Hearth and I concluded that Webb would quickly reconsider and that our first priority was to try to resurrect and sell a more modest program to land on Mars. In at least this instance we turned out to be correct.

Our first objective was to reinstate the Mariner Mars orbiting mission in 1971 to map the surface and help select future Mars landing sites. With the help of skillful maneuvering by Homer Newell this relatively inexpensive mission was quickly restored, and Homer also applied his good judgment to provide for two launches rather than one (thank God). We had started to regain a tiny bit of momentum.

Titan Mars '73 Is Born

We didn't have much hope for getting any kind of Mars lander in the fiscal year 1969 budget request, but I decided to give it a long-shot try. Much experience with simple hard landers had been gained at JPL and Ford Aeronutronic during development of the Ranger lunar capsules. Considerable design work had been performed on a scaled-up Mars version of the Ranger rough-lander capsule, and simulated Mars landings of this Mars capsule had been performed. (At Aeronutronic every time we blasted the 230-kilogram Mars capsule out of our air gun and into a soil box at more than 320 kilometers an hour we broke a few windows in nearby homes, which soon curtailed our testing.) Using this experience, plus the Surveyor cost data and all of the studies performed by JPL and contractors for Voyager, I worked over a weekend preparing charts showing the capability and estimated costs of various systems for landing on Mars. I estimated the cost of a Titan-launched orbiter with a modest hard lander at $364 million, compared to $600 million for an equivalent soft lander with a comparable basic payload of instruments.

Don Hearth and John Naugle thought the hard lander cost was low enough that it might just be considered, so we presented the story to Homer Newell, even though the fiscal year 1969 budget request had already been wrapped up with BOB. Now, Homer never told us what arguments he used with Jim Webb, or how Webb maneuvered with the White House, but our hard lander defied the long odds and somehow made it into the budget request titled as Titan Mars 1973. President Lyndon Johnson even gave it an enthusiastic boost in his 29 January 1968 budget message to Congress. To enhance our prospects with Congress, Webb soundly advised us to delay selecting a prime contractor for as long as possible to keep all the bidding companies lobbying for funding the project.

Throughout the several years of Voyager planning, JPL and Langley had been competing for overall management responsibility for the project. Those of us at NASA Headquarters had very ambitious plans for planetary exploration. We hoped to have many new missions, and we did not want to be limited by the size of the staff at JPL. For that reason, plus technical capability and a fine management performance on the Lunar Orbiter Project, Don Hearth assigned project management responsibility for Titan Mars '73 to NASA's Langley Research Center. James S. Martin, who had done an impressive job as deputy project manager for Lunar Orbiter, was appointed project manager for Titan Mars. We immediately gave Jim some of our precious Advanced Programs funding to get studies started.

The contractor teams from Martin Marietta, General Electric, Boeing, Hughes, and McDonnell Douglas with Ford Aeronutronic, which had been working on Voyager Mars proposals, immediately cranked up again on the 1973 Mars opportunity. They gave us a ton of supporting data with which we were able to impress the congressional staffers. Congress then readily approved the project as a much better bargain than Voyager and clearly not a disguised beginning of a man-to-Mars project, as had been suspected by some, including the aging but still influential Sen. Clinton Anderson.

The Evolution to Viking

In November 1968 Jim Martin announced that he was ready to give us the results of his Langley contractor studies on the 1973 Mars mission. Don Hearth, John Naugle, and I went to Langley for this very important briefing. A key participant at the meeting was Ed Cortright, who had recently been appointed director of the Langley Research Center. Jim Martin presented the study results and concluded that with judicious use of Apollo subsystems and hardware NASA could build a Mars soft lander for the same $364 million proposed for a hard lander. What? Now I believe in miracles, sometimes, but not this particular one. I counterargued, Don Hearth expressed only mild doubts, and an uncertain John Naugle turned to his trusted and wise friend Ed Cortright for guidance. I could sense the wheels turning in Ed's mind. I think he found the results doubtful, but he didn't want to start off as director of Langley by undermining his own newly acquired team. He didn't exactly say that he believed the results but rather made a noncommittal statement that he didn't see anything obviously wrong in the studies. John Naugle, like the rest of us, much preferred a soft lander to a hard lander if we could afford it. "Let's go for a soft lander," he said, leaping to the conclusion that it could be built within the $364 million budget. Meeting adjourned.

Now, every major space project requires a coordinated team effort, but right here I should tell you that the saga of this Mars mission, renamed Viking by the newly appointed NASA administrator, Thomas O. Paine, was dominated unlike any other NASA project by just one man—Jim Martin. Every successful planetary mission had a strong project manager, such as Pioneer's all-capable Charlie Hall, for example, but Jim was in a class by himself. If I make him out to sound like the villain of this story, be patient, he will indeed initially qualify for that role, but this saga will have many twists. In any event, he looked the part of the central character—a very large and powerful man, standing well over six feet tall with a solid square face and a close-cropped crew cut. A giant of a man in more ways than one.

My troubles with Jim began when we first started the Titan Mars studies at Langley. He came to me and protested the hard lander concept. He said he and Langley management wanted to manage a "first-class" project, not a cheapy hard lander. I agreed with him that a soft lander with a full complement of instruments was much preferred but that we were very fortunate to get even the more modest hard lander into the budget. Jim went away unconvinced. I'll never know whether he really believed those study results he presented in November 1968. I would like to believe that he did (I have since found him to be a difficult man, but honorable) and that he was influenced by the strong preference for a soft lander that we all shared.

To his credit Jim came back to us just a few months later in August 1969 to say that the Viking soft lander mission cost was going to be more than $600 million rather than the budgeted $364 million. We immediately requested a special hearing with our key congressional committee, the House Authorization Subcommittee on NASA Oversight chaired by the ever-tough Joe Karth. At the hearing John Naugle rounded off the cost estimate to what he hoped would be a comfortable $750 million. We took our licks at the hearing but got complimented off the record for our honesty in bringing the new estimates immediately to Congress. They acknowledged that we had given them the opportunity to cancel the project right then and there before we had spent any significant money. Amazingly, and with some credit possibly going to the fact that Ed Cortright was a frequent golfing partner with Karth, the more expensive Viking project stayed in the approved budget for NASA.

Our budget performance after that was quite respectable, but the budgeteers at the president's Office of Management and Budget had memories like elephants and for years kept reminding NASA of that jump in cost estimates. In my book it wasn't a case of cost growth, it was just that the $364 million figure had been pure fantasy.

A major contributor to successfully convincing skeptics such as Karth that we now understood the technical aspects and cost implications of the new Viking orbiter–soft lander mission was the contractor study effort, which far exceeded the level of effort paid for by our NASA funding. Al Kullas and Carlos de Mores of Martin Marietta in Denver were especially adroit at anticipating the data and charts that we would need to answer probing questions by the Space Science Board, the Lunar and Planetary Missions Board, and congressional staff members. It was not surprising that Martin Marietta later handily won the competition to become the Viking systems prime contractor.

Of the contending teams from Boeing, Martin Marietta, and McDonnell Douglas, the two front-runners appeared to me to be Martin Marietta and McDonnell Douglas. Both were very experienced and capable, so attention tended to focus more and more on their different approaches to the Mars landing vehicle. Landing tests for the Apollo lunar lander had shown that three-legged landers could easily tip over if the lander hit the surface with any appreciable lateral velocity. The landing multibeam radar would have to do a near perfect job to achieve a smooth landing. McDonnell Douglas got around this risk with a clever design that used a disc configuration for the bottom of the lander. Tests showed that this design could tolerate a large amount of lateral velocity at touch down. However, the proposal team at Martin Marietta knew that the engineers at NASA Langley, who would be evaluating the Viking Lander proposals, had done most of the simulated landing tests for Apollo and were comfortable with the three-legged configuration. Moreover, Langley was known to be rather conservative and the disc shape was "new." Martin Marietta proposed three legs and won the contract. There were of course many other factors in the proposal evaluations, but a three-legged configuration certainly did not hurt Martin's chances.

Getting any kind of scientific endorsement for Viking from NAS would have been much more difficult were it not for the strong support from Harry Hess, then chairman of NAS's Space Science Board. As a navy captain in World War II, geologist Hess had collected ocean-floor data that proved spreading across the Atlantic Ocean ridge and confirmed Alfred Wegener's earlier theory of continental drift. As this concept of tectonic plate movement was further verified by continental magnetic alignments, professional respect for Harry soared. On the SSB he steadfastly urged his fellow scientists to recognize that the Apollo Program was a venture in national prestige, not a science project, and that the science community should not fight it but get what science they could out of it and hope to ride its big-budget coattails in getting other science projects funded. Harry used some of the same arguments in defending Viking. Although Viking objectives might better be accomplished with several smaller missions, the White House and the NASA administrator liked the blockbuster approach, and it could indeed achieve the major science objectives for Mars, including a search for extraterrestrial life, one of the NAS's main objectives.

While the selling of Viking to the science community, the White House, and Congress was going well we were suddenly struck by a budget crisis from within. Fiscal year 1971 was to be the big spending year for Viking, requiring $212 million to make the 1973 launch date. As the administration's budget was being wrapped up in the final days of December 1969 we were told that Robert Mayo, the budget director, caught President Nixon as he was about to board his plane to San Clemente and advised him that he had been unable to meet his promised balanced budget by $7 billion. We were told that Nixon just snapped, "Well, balance it," and went on aboard Air Force One. NASA's share of the required cut was $225 million, with just twenty-four hours to revise its programs. There wasn't time to make many small cuts in numerous programs—the only viable solution was to cut a major program, and Viking was unfortunately the logical choice. Rather than cancel the hard-won project, however, it was slipped two years to the next Mars launch opportunity, 1975.[8] In retrospect and in view of the enormous technical challenges the project plunged into, that two-year slip was a godsend. At the time, though, we were faced with yet another cost increase due to the more difficult launch year and the cost of keeping teams intact.

Once again we faced Joe Karth and his committee with the news that the Viking estimated run out cost was now $830 million. It was a rough day, and Karth even chewed hard on his old buddy Ed Cortright, but Jim Martin and his teams at Langley, JPL, and Martin Marietta had done their homework in great detail and Karth and Congress eventually confirmed their support for the Viking Project.

In judging the cost performance of the managers of the Viking Project, I believe it is only fair to judge on the basis of this $830 million estimate. This was the first point at which the mission had been fully defined, including not only a soft lander but with the launch year set at

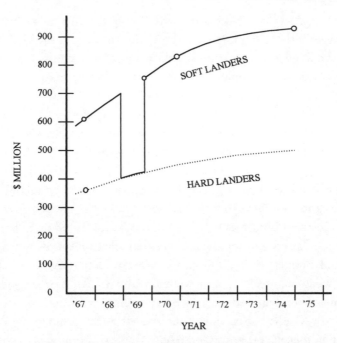

The so-called Wishing Well graph, showing the history
of NASA cost estimates for Mars landing missions and
illustrating the brief wishful-thinking period when NASA
hoped to perform a sophisticated soft landing on Mars
for the same price as a simpler hard landing.

1975. Every estimate prior to that was before the mission had been defined and before any great sums had been spent.

At about this time George Low, NASA's deputy administrator, organized a meeting of all of NASA's project managers to discuss how
to better estimate and control the cost of NASA's space projects. George
was driving for lower costs. I was cautioning against wishful thinking
instead of real cost reductions and displayed my "Wishing Well" curve,
a plot showing estimated cost curves for typical Mars landers, the
upper curve for soft landers and the lower curve for hard landers. In
November 1968 studies for soft landers showed them costing about
$600 million, with hard landers running at about the $364 million we
proposed for Titan Mars '73. We then entered the Wishing Well and

told Congress we could do a soft lander for the same price. Never true, just wishful thinking. We quickly corrected ourselves to a more proper, inflation-adjusted figure that Naugle rounded up to $750 million. Two years' delay to 1975 added more inflation, which contributed to boosting the estimate to $830 million.

Assignments to NASA Centers

With a significant portion of the NASA budget riding on this one project we needed strong and well-defined management responsibilities. Martin Marietta as contractor to Langley would build the Viking Mars Landers and integrate the entire assembly for launch; JPL was directed to do the Viking Mars Orbiter spacecraft in-house, using all their Mariner Mars Orbiter know-how but responsible to Langley as the overall mission manager; and a team at NASA's Lewis Research Center under Andy Stofan (who later was promoted to director of the center) would work with Martin Marietta to develop a new Titan 3–Centaur launch vehicle configuration. All of this, including science instruments from various universities and contractors, would be under the overall management of Jim Martin.

Jim needed capable help and he got it. I helped talk a highly respected exobiologist I had worked with at JPL on the ABL development, Gerald "Jerry" Soffen, into transferring to Langley to serve as Jim's project scientist. I knew that Jerry was not only a fine scientist but was a gentleman of the highest order, was people-oriented and smooth enough not to unduly antagonize Jim Martin, and would work well with the daunting array of Viking investigators, scientists at NASA Headquarters, and various review committees such as the SSB. A sensitive person, Jerry had a difficult time getting used to the authoritative and sometimes rough ways of "Big Jim" Martin. At NASA Headquarters the Viking program manager, Walt Jakobowski, the program scientist Dick Young, and I had to give Jerry occasional pep talks and encouragement to help him maintain his characteristic enthusiasm.

Ed Cortright at Langley was never able to meet all of Jim's sometimes staggering requests for support, but he assigned about three hundred

Langley engineers and scientists to Jim's Viking team. A number of these people had very solid experience in designing and managing the successful Lunar Orbiter Project. Senior among them was Israel "Iz" Taback, the former Spacecraft Systems manager from Lunar Orbiter. With the title of deputy project manager under Jim, Iz really served as the chief engineer throughout the Viking system design and development. He could be called the "father" of the Viking Lander design in the same sense that Max Faget was father of the space shuttle design. In addition to being highly respected for his strong technical judgment, Iz, who always had a ready smile and a twinkle in his eye, was one of the most universally liked members of the Viking team.

Among the younger recruits from Lunar Orbiter was A. Thomas Young, who started out as the very impressive Viking science integration manager and wound up as deputy to Jim Martin and mission director coordinating all of the complex and exciting operations of the Viking Landers on the surface of Mars. Tom's relaxed, southern-gentleman manner did not in any way obscure his obvious intelligence and communications skills. (We had to sell no one to have him later replace me as director of Lunar and Planetary Programs when I left NASA Headquarters. He then quickly advanced to become deputy director of NASA Ames, then director of NASA Goddard, and on to become president of Martin Marietta and president of the AIAA.) A talented guy, indeed, and a pleasure to work with.

Another Lunar Orbiter recruit was Cal Broome. Cal is of fairly light build and wears overlarge eyeglasses, so he tends at first glance to appear somewhat timid, but his deep voice and firm manner quickly dispel that impression. He started out on Viking coordinating all of the Lander science instrument definition and later concentrated on the Lander facsimile cameras to guide them through their difficult development. He earned the respect of the science team members, which is no mean feat.

Management from Headquarters

Before describing the Viking management team at NASA Headquarters, I need to give a bit of background on how Code SL functioned in the 1960s and 1970s. To strengthen and support a growing base of planetary scientists, SL had a team of senior scientists, each charged with building capability in an important specialty. For example, Bob Fellows supported a strong effort at many universities in Planetary Atmospheres, Steve Dwornik supported Planetary Geology, and Bill Brunk promoted Planetary Astronomy. Brunk's job was a bit unusual in that SL's main charter was to explore the solar system using robot spacecraft. However, it made economic sense to learn as much as possible about the solar system using Earth-based telescopes before we started to fund the development of expensive spacecraft. The problem was that the major telescopes were in great demand for viewing time, so that senior committees had to assign priorities to the various viewing requests. These committees gave little priority to astronomers who wanted to view our planets—they considered it much more important to study the origin and evolution of the entire universe rather than just the bodies orbiting around one very ordinary star. So SL had found it necessary to build and support telescopes dedicated to lunar and planetary observing, such as Prof. Gerald Kuiper's sixty-one-inch telescope in Arizona, the McDonald Observatory in western Texas, and the later Infrared Telescope Facility in Hawaii. Bill Brunk had the interesting job of promoting and managing this program.

As specific planetary spacecraft projects were approved, Headquarters program offices were established, using the word "program" to distinguish them from the "project" offices at the NASA centers. Typically an SL program office would be comprised of a program manager, an engineering deputy, a program scientist (who usually had other assignments as well), and a supporting secretary (very important to the success of the team). During the early and mid-1960s the challenge was just to make launch vehicles and spacecraft work, never mind the science. In those days the program managers were very much involved in the technical decision making. During the struggling development of the Surveyor lunar landing spacecraft, for example, Oran Nicks and Ben Milwitsky, the Surveyor program manager, were deep into the technical details. However, by the late 1960s the project managers at JPL, Ames, and Langley had both technical and management matters pretty much under control.

By the time I became director of SL in 1970 I believed that the program managers should not be trying to do day-to-day project management. From my experience of seventeen years managing a variety of R&D projects as a contractor, I believed strongly that the project manager should be held fully accountable for bringing his or her project to a successful conclusion on schedule and within budget. To do this the project manager needed to have the authority commensurate with the responsibility. I therefore believed that the primary role of NASA Headquarters was to sell the new projects and then assign them to strong project managers at the NASA centers, to interact with the science community to select the science instruments to be carried, to monitor progress to assure that the project was meeting schedule and budget, and then to do all that is necessary and prudent to defend that budget. Day-to-day technical decisions should be left entirely to the good judgment of the project managers.

By 1970 we had experienced planetary project managers at JPL, Langley, and Ames who had demonstrated their competence to manage difficult spacecraft projects. At Headquarters SL program managers such as Bill Cunningham, Earl Glahn, and Glenn Reiff were working very well with their project managers, were current on the status of their projects, and were supplying the necessary support from NASA Headquarters. This freed up the SL triumvirate of the director, deputy director (chief scientist), and manager of Advanced Programs to concentrate on their most important function, the planning and selling of new exploration projects to advance planetary science.

Viking did not quite fit comfortably into this pattern of program management. In the beginning I wasn't sure that I could trust Jim Martin to tell me the true story of progress and problems. Viking was so expensive that a mission failure would shake the entire agency down to its toes, and even a modest cost overrun, say, just 10 percent, could cause the cancellation of several smaller science projects. That would generate a science revolt against future planetary missions that could last for years. So we had to have good advance warning if Viking was heading into budget problems. I told SL's Viking program manager, Walt Jakobowski, that his primary job was to keep the lines of communication open with the Langley Project Office, even if that meant swallowing his pride and buttering up Big Jim at all times. The always-composed Walt had been appointed to his program manager position by Don Hearth after performing very well on prior programs such as Surveyor and was up to this challenge. We gave him two outstanding and experienced deputies in Bob Kennedy for the Orbiter and Rod Mills for the Lander (Rod had done an excellent job as head of my Mars ABL team at Aeronutronic and was persuaded to continue his Mars work with us at NASA). I told the Jakobowski-Kennedy-Mills team that I would be the bad guy whenever the need arose to argue with Jim. Our Headquarters Viking program scientist, Dick Young (former head of all exobiology and life science activities at NASA Ames) was to similarly stay in close touch with his friend Jerry Soffen, the Viking project scientist. These guys performed this delicate job marvelously well, and Headquarters was always fully informed in painful detail throughout the tortuous development of Viking.

At the time we started Viking our orders from BOB were to not acknowledge that inflation existed. This was ridiculous and guaranteed overruns. For Viking we were able to get an exception and budgeted for an inflation rate of 5 percent per year (we actually encountered double-digit years of up to 16 percent, and up to 25 percent within the aerospace industry). In addition we added in a necessary contingency reserve under the title of allowance for program adjustment (APA), a name thought up by Don Hearth and successfully defended by him as necessary to accommodate the inevitable increases in science scope as negotiations proceeded with the instrument principal investigators. When I took over from Don in 1970 my main financial management tool for the Viking Project was to track this contingency budget. I insisted every time the Project Office made a design decision affecting cost that any future lien must be subtracted from the APA reserve, even if the cost impact would not be felt for several years downstream. For example, if the Project Office early on in 1970 decided that a new test costing $1 million would have to be conducted at Cape Canaveral just before launch, they had to record that $1 million as an immediate subtraction from the APA reserve. Following the APA trend line indicating our remaining reserve funds gave me a pretty good indication of whether we were going to wind up staying within our Viking budget.

To his credit Big Jim never played games with this system and followed it in exact detail. I think he found it to be a valuable management tool for his own use. The APA system worked well and gave us timely warning of present and future cost problems.

Science on the Viking Lander was going to have to be done in a different manner. On a typical NASA science mission the PIs are selected through competition and are then responsible for supplying their instruments to be integrated onto the spacecraft. With Viking, however, the Lander was so constrained in both volume and weight that instrument design was going to have to be carefully coordinated to fill every available cubic centimeter. Moreover, organic content had to be strictly limited and everything had to withstand the rigors of heat sterilization. The Viking Project Office and its prime contractor, Martin Marietta, were clearly going to have to steer and control Lander instrument development.

Instead of individual PIs for Viking we assembled teams of scientists. Headquarters scientists Don Rea (science deputy under Don Hearth), Mike Mitz, and Dick Young worked closely with Jerry Soffen at Langley to select the most qualified scientists. The recommended teams were then approved by John Naugle. There was the usual tendency to keep adding experts to the teams, so that by launch time their number had grown to eighty. As leading scientists are by nature and necessity competitive and frequently self-centered, a management essential was to pick science team leaders who could draw their members together into productive units. Soffen et al. did a fine job in selecting the team leaders:

Science Team	Team Leader
Active Biology	Harold "Chuck" Klein
Lander Imaging	Thomas "Tim" Mutch
Molecular Analysis	Klaus Biemann
Entry Science	Alfred Nier
Meteorology	Seymour Hess
Radio Science	William Michael
Seismology	Don Anderson
Physical Properties	Richard Shorthill
Magnetic Properties	Robert Hargraves
Inorganic Chemistry	Priestly "Pete" Toulmin III

The team leader with the greatest challenge was Chuck Klein. Exobiologists proved to be especially strong individualists, and several continued to act strictly like PIs and not team members.

Ambitious Scientific Scope

True to his objective of making Viking a first-class mission, Jim Martin negotiated advanced features into the Viking science instruments. Don Rea and Dick Young at NASA Headquarters had guided the selection of science investigations for both the Orbiter and the Lander, but the design details were left to the project manager to fit within his weight and volume restraints, the set schedule, and his budget. Jim's pursuit of excellence, however, led to pushing the state of the art in many instruments and subsystems, so much so that the project was soon in budget and schedule problems.

We at Headquarters were aware of the extremely ambitious scope being established—why did we not object? Here we must reflect back on the space race with the Soviet Union. In spite of the official cold war, there was substantial scientific dialogue going on underground between scientists in the United States and their counterparts in Russia. We had no input into the future mission planning in the USSR, however. In truth, the Russian scientists knew little more than we did—with no warning they would get notice that one of their instruments had just been launched to Venus or Mars on a government-designed spacecraft. So all we at NASA had to go on was the history of and trends in the Soviet space program. We knew that the Russians had been persistent in attempting to land on Mars at every twenty-six-month launch opportunity; we had to assume that they would try again in 1971, 1973, 1975, and so on. To avoid being redundant, we had to surpass in scientific scope anything that their landings might achieve. Viking science had to be first class.

We found out later from sources such as Roald Sagdeev, former director of the Soviet Space Research Institute, that we had indeed put pressure on the Soviet space planners. They had quickly conceded the outer planets to the United States—with only an antenna-bearing ship to provide a tracking station outside the USSR, it would be very diffi-

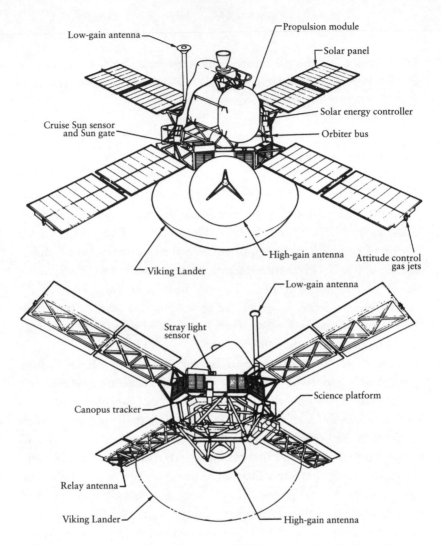

Low-gain antenna

Propulsion module

Solar panel

Cruise Sun sensor
and Sun gate

Solar energy controller

Orbiter bus

High-gain antenna

Attitude control
gas jets

Viking Lander

Low-gain antenna

Stray light
sensor

Science platform

Canopus tracker

Relay antenna

Viking Lander

High-gain antenna

Viking Orbiter, evolved from the successful Mariner 9 Mars Orbiter, shown with the attached Viking Lander capsule. (Courtesy NASA)

cult for them to adequately track and communicate over flight times of several years with far-distant spacecraft. The United States had a great asset in the worldwide Deep Space Network, with its large antennas in California, Australia, and South Africa (later moved to Spain), which had been built up so skillfully by Eberhardt Rechtin and his crew at JPL. So the Soviets decided to concentrate on nearby planets Venus and

Mars. Their assessment of the ambitious scope of Viking science was that it would surpass anything they could do at Mars during the 1975 opportunity, so they put great pressure on their team to achieve a successful Mars landing in 1973. More on that later.

Our emerging Viking schedule and budget problems were with the first-class Lander, not the more straightforward Orbiter. JPL had negotiated a comfortable Orbiter budget with Jim (he was later able to take away $16 million to help pay for the Landers) and were rather conservative in their approach of scaling up their Mariner 9 Mars Orbiter with known advances in imaging and communications.

As an example of a first-class Lander instrument, let us consider the imaging camera. A slow-scanning facsimile camera, essentially just a scanning light meter, had been developed at Ford Aeronutronic for the Ranger Lunar Program. It produced beautiful pictures and was quite suitable for use on Mars. For Jim Martin's Viking Lander, however, two cameras were specified to give stereo depth perception, and then full-color capability was added. And not just full color but also several infrared wavelengths. All this to be sterilized for twenty-four hours at over 250 degrees Fahrenheit, of course. It is not surprising that camera development ran into difficulties.

For organics detection an entirely new instrument was to be developed. The most basic instrument for analysis of soil constituents was to be a very sensitive mass spectrometer, but mass spectrometers are easily saturated and work best when fed just one gaseous compound at a time. The separation of a mixture of gases could be done with a gas chromatograph, a laboratory diagnostic instrument that runs gaseous mixtures through a filtered column that lets lighter gases through first and slows down those with larger molecules. For Viking a combination gas chromatograph–mass spectrometer (GC-MS) was specified that would be sensitive down to just one part per billion and cover molecular ranges up to two hundred atomic mass units. All of this in a package the size of a portable typewriter. Within that tight volume it had to include its own soil processor to grind up a soil sample and then heat it in stages up to five hundred degrees Celsius to vaporize any interesting organic compounds.

It is not surprising that first JPL and then a contractor team of Litton Industries, Perkin Elmer, and Beckman Instruments had serious troubles developing that breakthrough GC-MS instrument. It is in dealing with trouble, though, that good people emerge. For example, the Langley person who wound up responsible for pushing the GC-MS was a very young engineer by the name of Al Diaz. Al delivered, and he progressed rapidly after his Viking days, serving ably at NASA Headquarters as deputy associate administrator for Space Science and then as director of NASA's Goddard Space Flight Center.

The most ambitious and challenging instrument of all was the life-detection package. First of all, what is "life"? There is no precise definition. All biologists can do is list certain properties that known life-forms possess. And yet some organisms, such as viruses, only exhibit some of the normal characteristics of living organisms. Among these common characteristics are the ability to draw energy from nutrients or solar radiation and the ability to replicate, that is, under favorable conditions to increase and multiply. The Viking biology instrument focused on these two properties of life and wound up being an integrated assembly of four separate experiments—the pyrolytic release experiment of Norman Horowitz, the labeled release experiment of Gil Levin, the gas exchange experiment of Vance Oyama, and the light-scattering experiment of Wolf Vishniac. Together the four provided a range of environments from very dry to saturated solutions.

It was soon apparent that we had bitten off more than we could chew. On Earth these four experiments would require a large laboratory crammed with equipment; on the Viking Lander we were allotting just one cubic foot, little more than the size of a gallon milk carton. Into that cube must go multiple ovens, many ampules of nutrients, radioactive-gas bottles, a solar simulator, many dozens of valves, and a computer containing twenty thousand transistors to adaptively program successive experiment procedures should the Lander lose its uplink commands from Earth.

Challenging Hurdles in Technology

By early 1972 it was becoming apparent to the biology instrument engineers, if not to all of the team biologists, that we just were not going

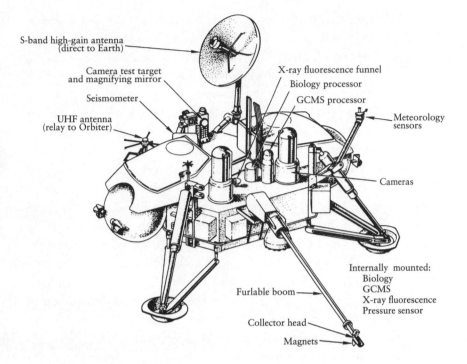

S-band high-gain antenna
(direct to Earth)

Camera test target
and magnifying mirror

Seismometer

UHF antenna
(relay to Orbiter)

X-ray fluorescence funnel

Biology processor

GCMS processor

Meteorology
sensors

Cameras

Internally mounted:
Biology
GCMS
X-ray fluorescence
Pressure sensor

Furlable boom

Collector head

Magnets

Viking Lander, RTG-powered and carrying a dense array of scientific
instruments and an extendable boom for collecting soil samples from the
Martian surface. Viking 1 made the first successful landing on another planet
on 20 July 1976. (Courtesy NASA)

to make it with four life-detection experiments. We seriously considered cutting all the way down to just one experiment, but that, we decided, would greatly compromise any conclusion on the presence of
life, so we (optimistically) concluded we could squeeze three experiments into the one-cubic-foot package. We appointed three members
of the Biology Team who were not affiliated with any particular experiment (Chuck Klein, Nobel laureate Josh Lederberg, and Alex Rich)
to recommend which of the four experiments to drop.

Any cancellation was going to be traumatic, but it was especially
painful when they picked the light-scattering experiment of Wolf Vishniac. Of all the team members Wolf stood out as the one who was always enthusiastic, cheerful, and campaigning harder than any other bi-

ologist for the new field of exobiology. It was personally distressing to me, as he may well have saved my life in 1970 in Leningrad when I developed internal bleeding after an accident. It was Russian-speaking Wolf who managed to get an ambulance out of nowhere and never left my side until convinced that the Russian doctors had everything under control. Of all people to pick for cancellation, why Wolf? Well, the answer was that his experiment was conducted with lots of liquid water, the least Mars-like environment. Couldn't really argue with that.

Wolf was devastated but nevertheless stayed on to contribute to the team effort. In December 1973 he traveled to Antarctica to study whether microbes could survive in the most Mars-like dry valleys found on Earth. While out hiking alone on a steep slope in one of the dry valleys he slipped and fell to his death. I promised his widow that if we found life on Mars I would campaign to have the organism named after Wolf. Viking did not find life, so instead, after Mariner 10 viewed Mercury in its 1974 flybys, we got one of the craters on Mercury named after Wolf. But we may still find microbes on Mars and, if so, I intend to do my best to keep my promise to Mrs. Vishniac.

Even with just three experiments the biology instrument was presenting enormous difficulties. The instrument contractor, TRW, had reduced the size of the instrument plumbing down to capillary tubes with passages about the size of a human hair. The valves were so small they were hard to see. The instrument was almost impossible for humans to assemble, much less to expect it to work. Jim Fletcher, when he came aboard as NASA administrator, would catch me in the NASA lunch room at least once a week and ask me how we were doing on Viking (this in addition to the formal reports he was getting). He encouraged me to find some simpler "litmus-paper" test for life—some magic paper you could dip into a solution and it would turn red if there were living organisms present. Wishful thinking. Unfortunately, no such paper exists. Biology was not Fletcher's field of expertise, and he had not been exposed to the years of debate and discussion of the difficulty of detecting unknown forms of microorganisms.

With only months to go before the instrument was due to be delivered we appointed a review "tiger team" headed by JPL's instrument specialist Dave Rogers along with JPL's Kane Casani, a respected engineer I had worked with on the Ranger Project. The review team soberly concluded that the one-cubic-foot bio instrument was "as complex as most complete spacecraft" and that we would never make delivery in time for launch. In other words, give up.

Part of the problem was that the package was so small that only one person at a time using miniature dentist's tools could work on the assembly. You couldn't speed up the work by assigning more people. TRW's instrument manager Gene Noneman and his science deputy Fred Brown tried using women technicians with smaller hands and nimble fingers, but they lacked the necessary mechanical experience. I tried any motivation I could think of—once in a review meeting when Hatch Wroton, the Martin Marietta resident engineer at TRW, admired my necktie I immediately took it off and gave it to him. Any little thing for a guy who had left his family for months and was working twelve hours a day, seven days a week.

Meanwhile, those of us concerned with the budget, including John Naugle and Martin Marietta management all the way up to board chairman Tom Pownall, were hammering on George Solomon, TRW's general manager, about cost, schedule, and management. TRW's final cost more than quadrupled the original contracted price. Through dedicated determined effort and a lavish use of overtime, though, the biology instruments were delivered in good condition in time for launch. And the instruments actually worked! To TRW's credit I can't really think of any other contractor who could have done better on this almost impossible task, and I doubt if there are any who could have done as well.

Meanwhile, engineering subsystems for the Viking Lander were also in deep trouble. Again Viking was pushing the state of the art on many fronts. Martin Marietta was on a partially cost incentive contract and thought they had been clever in negotiating fixed-price contracts with most of their subcontractors, but some of these subcontractors, when they foresaw a probable cost overrun, began to take workers off the effort in a desperate attempt to avoid a loss. This put delivery schedules in increasing jeopardy. The Viking Project had a fixed launch date and couldn't afford schedule slips. With some of these subcontractors we had to swallow our sense of righteousness and renegotiate their contracts from fixed price to award fee in order to get them to increase effort to get back on schedule.

One such subcontract was for the Lander's central computer, being developed by Honeywell in Florida. The current state of the art for the computer memory was to use graphite cores, tiny doughnuts of carbon through which a matrix of wires was woven. But Honeywell had demonstrated success using a matrix of just plated wires of only .005-inch diameter to produce a smaller and lighter memory package. Moreover, they were willing to commit to reduce that further to .002-inch wire and still sign a firm fixed-price contract. Well, .002-inch wire, the size of a fine human hair, proved to be a different kettle of fish. Even skilled women technicians couldn't thread it through the matrix without kinking it, which ruined its magnetic properties. Special tooling could do it, but Honeywell management didn't want to spend the money under its fixed-price budget. We at NASA joined with Martin Marietta to put enough pressure on Honeywell to get a new plant manager assigned, and we reluctantly rewrote the contract to give them restored incentive to perform.

The plated-wire memories were so late that the project had to start development of a minimum-capacity graphite core design as an insurance backup. The flight Landers couldn't even be tested because they were fully automated and wouldn't function without their onboard computer. Our Viking Project was in deep trouble. However, the ladies at Honeywell finally did the impossible and sent two good plated-wire units to the cape just in time to test the Landers before final launch preparations.

Some seemingly sound engineering decisions backfired. For example, each instrument and subsystem required a substantial complement of electronics. Several of the instruments, such as biology, required quite extensive computer capability to permit autonomously controlled successive experiment runs with varying parameters based on the results of prior sample runs. In other words, they needed their own brains (computers) to make decisions if they weren't able to receive instructions from their experimenters back on Earth. Even instruments that needed no built-in computers required sophisticated signal processors. To maximize reliability and contain costs the Project Office decided to require everyone to use a standardized hermetically sealed microprocessor chip known as a 54L. The project ordered a buy of 125,000 of these 54Ls to be fabricated and distributed to all the subcontractors. These chips were made under the tightest quality controls under laminar-flow hoods within super clean rooms.

During testing of the 54Ls one of them failed. When the hermetically sealed "can" over the microcircuit chip was opened, a particle was found to have shorted a circuit. In a microcircuit even a particle as small as .001 inch can interconnect two conductors and produce a short. Did other 54L cans contain particles? A shaker table with sensitive microphones was designed to shake the cans and listen for particles rattling around inside. Some cans with rattles were found and rejected. Then some of the "good" cans that had passed the test were opened. Particles were found inside! Did they somehow get introduced in opening the cans?

All kinds of techniques were tried with uncertain results. Finally Jim Martin decided to play it safe and order another 125,000 of the 54Ls but this time have them "glassivated," that is, coated with a layer of glass so that no particles could short out the circuits. Problem solved? No way. Corrosion was found forming underneath the glass layer. The chips didn't fail, but they might in time. We had run out of time—the subcontractors needed to proceed with assembling their flight units. The decision was made to use both the sealed and the glassivated 54Ls. Each unit on the Viking Lander was required to have redundant circuits, so sealed 54Ls would be used in one circuit and glassivated 54Ls would be used in the other. Hopefully one would work through the duration of the mission.

Actually, not one of the 54Ls failed during the Viking missions, and the Viking 1 Lander stayed in operation for years in the severe Martian environment, returning daily Mars weather reports. All of the thousands of leftover 54Ls were offered to other project managers, who were only too happy to receive free parts with a proven flight history. I never heard of one that failed in flight.

Rough Times for Management

As the Viking Lander ran into serious development and budget problems Martin Marietta made the typical corporate move—they replaced their project manager, Al Kullas. Now this undeniably has advantages for the contractor. It effectively wipes the slate clean. The new man can say that all prior mistakes were made by his predecessor and will not be made in the future. I wrote a strong letter defending Kullas, who had worked hard and effectively. The problems weren't due to his performance or lack thereof. But Al had been getting into increasingly deeper arguments with Jim Martin over the management of budget reserves, and so he was gone. (I used budget reserve as my chief financial control, and so did Jim. Al wanted to get out from under some of that control.)

Al's replacement, Walt Lowrie, got along much better with Jim Martin. Walt was a capable and experienced manager with strong technical judgment and, most important, the composure to stand up under the intense pressure that was to come in getting Viking to its launch pad. Walt had a strong team behind him. Especially outstanding was John Goodlette, the Lander systems engineer and perhaps the only person to really understand how the entire complex Viking Lander really worked.

Another standout for other reasons was Gentry Lee. Young, with hippy-long hair, Gentry had almost uncontained enthusiasm and energy and could talk at great speed for hours in the then-new language of computer software. We figured he was either a genius or a complete screwball. Nobody else we knew was on the cutting edge of this new technology, so we gambled on genius and put Gentry in charge of all software preparation. Fortunately, genius was pretty close to right; the software worked, and JPL later recruited Gentry to work for them to make projects such as Voyager and Galileo fulfill their complex missions. With a bit of his excess energy Gentry also served as the producer of Carl Sagan's *Cosmos* television series, which set all-time viewing records for a television documentary series.

By 1974 technical problems on the Viking Lander were getting so bad that, unbeknownst to me, Jim Fletcher with George Low and John

Naugle seriously debated replacing Jim Martin as Viking project manager. That would have been a fatal mistake. The problem was the overwhelming scope of the Lander, not Jim's organization or management. Fortunately they decided he should remain as project manager. There had already been enough sacrificial changes at Martin Marietta.

In the final crucial year of Viking Lander development Larry Adams, who had replaced Bill Purdy as general manager of the Martin Marietta Denver Division, gave Walt Lowrie further help by asking Dewey Rinehart to serve as Walt's deputy. Dewey was a Martin Marietta veteran who had built and managed the growing and successful Titan launch vehicle program. He was senior to Walt, and it was generous of him to willingly step down into a deputy role to help with a project in trouble. About that same time at NASA Headquarters John Naugle was promoted to NASA chief scientist (soon to be further elevated to associate administrator) and was replaced by the very talented Noel Hinners, an old friend of mine since he first came to Washington in 1962 to provide geological expertise to the Apollo Program. Noel was very strong on science and management but without direct hardware experience, so he brought onboard as his deputy John Thole, who had managed the successful Applications Technology Satellite (ATS) project at NASA Goddard.

Thole immediately was informed of the problems the Viking Lander was facing and, never lacking in self-confidence, decided to visit Denver to tell Martin Marietta how to solve all their problems. After a day-long meeting, with a cocky John and his Goddard experts doing most of the talking, Dewey was assigned to take the NASA visitors out to dinner. Soon after sitting down at the restaurant, and with Thole just warming up again on all his solutions, Dewey got up abruptly and said that Martin Marietta would pay the restaurant bill but that Thole was the most rude and arrogant SOB he had ever met and he refused to sit with him any longer. The next morning he phoned in his resignation from Martin Marietta. It took Larry Adams three days to talk him into coming back to work.

Thole was astonished. He didn't realize that his know-it-all, abrasive project manager style did not fit well with his new Headquarters role. He had come across like a bull in a china shop. Noel quickly found a new deputy. Indeed, people problems can frequently be just as disruptive as technical problems.

Noel's next deputy, Anthony Calio, took quite a different position on Viking. Tony was intelligent, extremely hard working, and just as extremely ambitious. He believed Viking was headed for disaster and therefore kept enough of an arm's length from Viking management that he could not in any way be blamed for its failure. As I said, Tony was very intelligent.

During these troubled times we "upper management" types were trying to help by getting the full attention of upper management at the subcontractors who were in trouble. To expedite our visits to "subs" we obtained legal concurrence for NASA managers to fly on Martin Marietta's very speedy Gulfstream 3 corporate jet. A great help. For example, early one morning I joined up with Martin Marietta president (soon to be chairman and hero-of-industry from fighting off a hostile takeover bid by Bill Agee) Tom Pownall at Washington's Dulles Airport. We flew south to Langley and picked up Ed Cortright and Jim Martin. Then on to Denver to get Walt Lowrie. While munching a sandwich and getting briefed in-flight by Walt about problems on the Lander propulsion system we flew to the responsible subcontractor, Rocket Research, located near Seattle. We had a brisk two-hour meeting with Rocket Research management, who were very responsive and asked for no relief from their fixed price contract, bless their hearts. Back to our plane at the local airfield, plane rolling even before the door was closed, and another briefing in the air on all the problems at TRW while we flew to Los Angeles. Two hours of beating on George Solomon at TRW, then on south to San Diego. Another briefing over dinner on the Lander radar development at Ryan, followed by a two-hour evening meeting with the Ryan people. All of this in one day, from D.C. to Virginia to Colorado to Washington to Los Angeles and almost to the Mexican border. Amazing how much territory you can cover when your plane is waiting for you at the nearest small airport and the plane starts rolling before you are even seated. Almost fun until the pilots inadvertently fly into a large thunderhead and suddenly you realize what a small craft you are riding in. Bouncy or not, we used this speedy aircraft to make similar whirlwind visits to ITEK, Hamilton Standard, Lockheed Electronics, RCA, Bendix, and Honeywell.

Russian Input

We had too many troubles of our own to worry much about beating the Soviet Union to the surface of Mars. Much later we learned from Roald Sagdeev that the Soviets were desperate to beat Viking to Mars. They had no instruments to match the sophistication of those on the Viking Landers so that anything they did after Viking would be overshadowed. At the last launch opportunity before the 1975 Viking window, in August 1973, they launched two spacecraft, Mars 6 and 7, to fly by Mars and deploy Landers to the surface. The Mars 7 Lander missed the planet, but Mars 6 was making a normal descent until it went off the air just above the surface when its final landing rockets were supposed to fire.

In an unpublicized circumvention of the cold war the Russian Mars project manager, Khodarov, came to Washington to visit us on 6 May 1974 and gave us a detailed briefing on everything they knew about the Mars 6 mission, what data it had returned, and why they thought it had failed. I am convinced he held back nothing. What interested us the most were the data transmitted from their mass spectrometer during the descent through the Martian lower atmosphere. It had failed to pump down to the expected level of vacuum. They believed this was due to an unexpectedly high level of argon, a noble gas that could not be handled by their ion pump. If this was true our Viking GC-MS was not going to work either. We rushed off to modify the operation of our instruments to allow our pump to handle the argon.

As it turned out, the Russians were wrong about their argon hypothesis. Viking found very little argon on Mars. However, this is still a good example of the considerable, albeit selective, degree of U.S.-USSR cooperation that was going on in planetary programs behind the scenes, with full knowledge of the State Department but under its directive to keep it out of the diplomatic spotlight.

Cost Concerns

During the peak spending years of Viking development inflation soared to as high as 16 percent, and in the aerospace industry it went over 20 percent. Our APA reserve was just not going to be able to handle this unexpected level of inflation on top of all our technical problems. Thanks to the good rapport Code SL's Viking Program staff of Jakobowski, Kennedy, Mills, and Young had with their counterparts at Langley, we were able to keep NASA top management as well as OMB and the congressional committees fully informed of our problems and what we were doing to solve them. Whenever the congressmen wanted we flew them down to Langley where Jim Martin and his team gave them an honest and candid assessment of our status, including a realistic projection of future costs. During the really tough times I went as often as every two weeks to the Hill to keep key congressional staff members fully up to speed on Viking progress. Jim Gehrig, Glen Wilson, and Craig Vorhees in the Senate were especially interested in details. The committees trusted us to tell it like it was, and I believe that contributed strongly to the excellent support we received from both the White House and Congress.

When it became pretty certain that we were facing an overrun in total cost I informed my boss, John Naugle, and Jim Fletcher's office what cuts in mission scope we would have to take to hold to our $830 million estimate. Jim Martin didn't like any cuts in the ambitious scope of his project. Twice when my options were getting serious consideration he and Ed Cortright came to Headquarters to present a rebuttal. The key decision maker each time was George Low, then deputy administrator, acting for Jim Fletcher. George believed strongly that NASA needed a major success in the mid-1970s. Apollo Soyuz was the only other major mission scheduled, and George didn't think the American public would get very excited about a handshake in space, so he wanted Viking to be, in his words, "a first-class mission." No cuts in scope. OK. I felt I had done my duty in warning top management of a near-certain overrun. The administrator's office had been given a realistic assessment and had made a fully informed decision. Years later, in July 1996 during the twentieth-anniversary celebration of Viking's landing, Jim Martin said that in effect he really reported to the NASA administrator. I always felt that he believed that, and to my chagrin, in many ways he truly did.

Things changed in March 1974 when Rocco Petrone came to Head-quarters for a brief year as associate administrator. Rocco was a pow-erful man, much like Big Jim Martin in many ways, and had been a major player in the Apollo Program and director of NASA's Marshall Space Flight Center. I had to give Rocco weekly presentations on the status of Viking, and he wanted all the details. At one review I had to tell him that Jim Martin's insistence on assembling a third spare lander in Denver was, in my opinion, stretching Martin Marietta's technician team too thin and jeopardizing the schedule for the two flight Landers. Rocco agreed. Jim and Ed tried the Headquarters appeal again. No dice this time; Rocco had spoken. I didn't like this chain of command (Jim and I should have been able to agree, and normally I wouldn't try to overrule a project manager's judgment), but I believe it was the right decision.

John Naugle tells me that he had great hopes at the time that Rocco would somehow get Viking out of its technical woes. Instead, Rocco told him that he found the problems "staggering" and beyond easy so-lution. Soon Rocco found that his talents were redundant with those of George Low and he departed NASA Headquarters.

Obviously these were times of great stress for managers on the Viking Project. As another spacecraft project manager in Europe, Richard Blott, said, "People ask me how I can sleep managing experi-mental satellite programs. I tell them I sleep like a baby. I sleep for an hour, then I cry for an hour, and I spend the rest of the night on the bottle."[9] I suspect that Walt Lowrie spent many a restless night. But I doubt that Jim Martin tossed much in his sleep. I think Jim was always confident that he could overwhelm all obstacles with sheer strength of effort and will.

I took pride in having no overruns in Planetary Programs and never handing my boss, John Naugle, a budget problem. (Maybe that is why we are still good friends.) But there was no way I could cover the

Viking overruns. Fortunately, Bill Lilly, the shrewd and tough-minded NASA comptroller, never believed that Viking could be done within its allotted budget. He managed to squirrel away some reserve funding within the overall NASA budget (no way even Bill could do that today) to cover the Viking overruns within Naugle's Space Science budget. Thanks to Bill Lilly's foresight Viking can say that it never caused the cancellation of any other large or small space science missions and programs. However, one could argue that its overruns might have delayed the start of potential new projects.

Up and Away

Through the outstanding efforts of hundreds of dedicated people the two Viking spacecraft (each an integrated Orbiter-Lander combination) and their Titan 3–Centaur launch vehicles made it to Cape Canaveral in time for checkout, assembly, and prep for launch. There were anxious moments with the spacecraft but none that required the dreaded opening of the Lander capsules that would have required a repeat of the sterilization cycle. If anything, there were teething problems with the new launch vehicles. NASA manuals designated the project manager as the person responsible for all phases of his mission, but in truth, launches were always left up to the launch experts. Not so with Jim Martin. He chaired all the major decision-making meetings at the cape on the Titan 3–Centaur as well as on the spacecraft. Actually, he and Andy Stofan functioned quite well as a team working with the Kennedy Space Center (KSC) and air force Titan people.

Some insight into the prelaunch preparations might be gained from a letter I wrote to my wife Anne from the cape on 14 August 1975: "Unfortunately we are having a few problems. First it was a bad valve on the launch vehicle, and then we had something (lightning?) inadvertently turn on our spacecraft and run down the battery. We are currently swapping spacecraft out on the launch pad. The whole process gets slowed down when we have to clear the pad during lightning storms. Tuesday afternoon we were held off the pad for six hours! With good weather it looks like our *earliest* launch dates will be August 20 and August 31, but nothing ever seems to come easily to the Viking Project. Just say a few prayers that we have no *new* problems."

Viking 1 got off to a clean launch pretty much on schedule on 20 August 1975. Viking 2 had scarier weather to deal with. Launches to Mars need to accelerate from Earth to go out from the sun and therefore are typically done in the late afternoon or early evening to take full advantage of the speed of Earth's rotation and movement around the sun. But in August and September late afternoon is just when almost daily thunderstorms occur on Florida's east coast. On 9 September, as we approached the opening of the launch window, I got up from my seat in Mission Control and ran outside every few minutes to check a growing mass of black clouds moving in from the west. Our official cape meteorologist was a young air force lieutenant, boyish looking but very sharp and decisive. After checking his data seven minutes before the opening of the launch window he announced that he was "go" for launch if we could get it off in the next ten minutes. It is probably good that Big Jim did not look outside, as I did, a few minutes later to see this huge black thunderhead towering almost over us. Jim trusted his meteorologist and coolly gave the final go. Photos of the liftoff show such a black background that most people assume it was a late-night launch. Just minutes after liftoff our television-camera views of the launch pad went completely blurred as rain let go with hurricane ferocity. But our Vikings were on their way to Mars.

Landing on Mars

During the Viking launch preparations at the cape I had gotten better acquainted with Hugh Harris, the energetic head of Public Affairs at KSC. He was impressed with the growing interest of the news media in Viking and offered to help me get published some articles I had written on my favorite sales theme, comparative planetology—how the study of Mars and Venus could provide invaluable insight into atmospheric processes on Earth (weather, climate changes, atmospheric evolution, ozone depletion, global warming, and the effects of atmospheric contaminants such as smoke, dust, smog, etc.). Using Hugh's contacts we were able to publish in thirty-seven different Sunday supplement magazines a color-illustrated article under various titles, such as "Mars: Tomorrow's Earth?" and "Planets Reflect Earth's Future," and with descriptive blurbs such as "In an age of pollution, poverty and energy shortages, traipsing among the planets may help man solve his problems." A few of these articles, which came out in May and June 1976, are listed in the references. The following is an excerpt:

> Accurate weather prediction can save billions of dollars annually for farmers and businessmen, avert famine in many parts of the world and help alleviate untold suffering and death. Predicting the weather, however, is like baking a cake. To achieve the desired result, you must choose the proper ingredients.
>
> One ingredient was discovered on Mars by a robot spacecraft. Mariner 4, which reached the red planet in 1965, revealed that heat radiation played a key role in dictating Martian weather. This factor had always been neglected in weather models on Earth. When it was added, weather prediction improved. Oceanless Mars also has provided us with a better understanding of the total weather spectrum by allowing us to observe atmospheric patterns and shifts unaffected by large bodies of water.
>
> But what about climate and long-term weather? If the theories of planetary evolution are correct, Earth will eventually become a Venus-like sphere enveloped by a mantle of poisonous gases. Here, though, we are talking about evolutionary processes measured in billions of years. But unlike Venus, Earth is being altered by man and his polluting machines.
>
> For example, each year millions of tons of smoke, dust and chemicals are dumped into Earth's atmosphere from manmade as well as natural sources. Will this increase Earth's average temperature by trapping solar heat and speeding the evolutionary heating process? Or will these pollutants block heat from the Sun, cooling Earth?
>
> Again, we must turn to another world for the answer. By a stroke of fate Mariner 9 slid into orbit around Mars in late 1971 at the height of a monstrous dust storm that completely obscured the planet's surface for a month. As the Martian skies cleared, Mariner 9's instruments continually recorded temperature profiles through the atmosphere to the surface.
>
> The result? The dust caused a drop in Martian surface temperatures of more than 40 degrees F., dramatic proof that similar Earth atmospheric pollutants could indeed trigger a disastrous cooling trend, or, worse, another ice age."[10]

Comparative planetology was a pretty new subject then, but today, thanks to the superior communication skills of people such as Carl Sagan and Lou Friedman of the Planetary Society, its value is widely recognized.

President Nixon had been told that Viking 1 could land on Mars on 4 July 1976, the two-hundredth birthday of the nation. Jim Fletcher immediately received a letter from the White House saying that in any future NASA program cutbacks Viking should not be touched. I believe the president already had this great achievement written into his planned bicentennial speech. As the Viking spacecraft circled Mars in June 1976, however, Jim Martin became cautious and decided he didn't like what he saw from the Viking 1 Orbiter's television cameras, so the landing was delayed for study of more Orbiter images and Earth-based radar data to increase the probability of a safe landing. I am pleased to confirm that neither I nor higher management at NASA Headquarters put any pressure on Jim to add any risk to the mission in order to make a "political" landing on the Fourth. After developing confidence that the Martian plains of Chryse looked like smooth sand deposited by the great floods that once formed the Valles Marineris, Jim gave the OK to kick the Lander out of orbit on 20 July.

As the Lander descended toward the surface I thought that I was very calm, was ready to accept a failure on the first landing attempt, and was already mentally rehearsing my pitch to finish assembling the spare third Lander for an additional backup launch in 1977. When the first radio signal came in to JPL to confirm that the Lander had survived on the surface I found that I had leaped from my chair at least a foot into the air. Who says white men can't jump? I had really been wound tight as a spring. Everyone in the mission control center was backslapping and celebrating in high style. Big Jim was trying to look dignified, but he was grinning from ear to ear. I even saw Jim Fletcher, a faithful Mormon teetotaler, sipping some champagne. Bruce Murray, one of the real hero scientists from the Mariner 10 Venus-Mercury mission and always active in the debate about the evolution of Mars, was fully immersed in celebrating the first landing on Mars in his newly appointed role as director of JPL.[11]

As the scanning pictures from the Martian surface came slowly line-by-line onto our screens, I began collecting bets. I never believed any place on cratered, erupted, and eroded Mars would be very smooth and had made bets that we would see rocks more than one foot in diameter. And there they were. Viking was lucky not to land on one of the larger ones. A camera view down at one of the Lander's foot pads showed that it had been fortunate to contact the surface on a relatively smooth sandy spot.

Early Results

As usual, we pressured our science teams to release some preliminary results within twenty-four hours, even at the risk of having to make later modifications to first impressions. The Lander Imaging Team released a first panoramic view of the Martian landscape, showing a rugged, reddish, desertlike terrain and a light blue sky, quite Earth-like overall. Within just a few hours they were able to better calibrate their Lander cameras and presented a corrected image showing the Martian sky to actually have a reddish tint, closer to pink than to blue. Not quite so Earth-like after all, which somehow seemed more appropriate for a different world.

The first few weeks of Viking operations on the Martian surface were enormously exciting as we viewed a rugged terrain from first Viking 1 and then the more northern and even more rocky landing site of Viking 2. The remotely controlled arms of both landers were busily scooping up samples of soil for analysis, including tipping over rocks to scoop samples that were more sheltered from the Martian atmosphere and from solar radiation. Author James Michener, who had come to Pasadena before the landing to join Norman Cousins, Jacques Cousteau, Ray Bradbury, and Philip Morrison on a panel to philosophize on "Why Man Explores" and on the potential impact of finding extraterrestrial life on Mars, was so fascinated by man's ability to reach out robotically across the solar system to dig at will in a new sandbox that he stayed on for an extra week to watch each new image come into JPL. My wife Anne was able to join him as he toured through JPL's image processing lab. Michener was inspired there to write his novel called *Space,* and he generously donated the royalties to establish a college scholarship fund for the children of NASA employees. (I was proud to be on the fund's first board of directors, where I was amazed to find out how many smart NASA kids we had who scored perfect 1600s on their SAT tests).

The Search for Life

Initially there were very active reactions from two of the three Viking life-detection experiments, the gas exchange experiment and the labeled release experiment, which could be interpreted as indicating the presence of active organisms. To their credit, Chuck Klein and his Active Biology Team did not try to make headlines by immediately announcing that they had found life on Mars. That had to have been a great temptation. Yet they were properly cautious in their interpretation— the experiment reactions just appeared to be too vigorous to be due to biological activity. Then data kept coming in from more and more samples that had been processed through the GC-MS instrument. No interesting organic compounds detected. Not even any simple organics. Nothing that would indicate life-forms or even organics from which life might evolve. The GC-MS instrument was working perfectly and, as you may remember, it could detect molecules from the simple to the fairly complex (200 AMU) with a sensitivity of one part per billion. There were just no organics on the surface of Mars, at least at our two landing sites. Great science, but a letdown. It would have been sensational to find life.

The final conclusion of the science teams and our project scientist, Jerry Soffen, was that the penetration of ultraviolet rays from the sun through the thin Martian atmosphere was so intense that it would not only sterilize the surface but would break down molecules of any complexity. The strong UV could create strongly oxidizing hydroxyl radicals on the surface that would explain the active reactions from the two Viking biology experiments. To this day Gil Levin does not accept that explanation for the results of his labeled release experiment and believes he detected organisms, but the consensus is strongly against him. However, nothing in the Viking data would rule out living organisms at water-ice boundaries deeper below the surface or at the edge of the polar caps. The story of Mars is far from complete.

The news media people swarmed over JPL during the early days of Viking operations on Mars. They quickly latched on to favorite spokesmen for Viking. Tom Young was eloquent, clearly describing Viking mission operations. Bushy-browed Mike Carr described the television views from the Orbiter, and Hal Masursky of the USGS did a great job interpreting and explaining the geology of the Martian surface. And, as usual, Carl Sagan was a news media favorite, explaining all of the Viking results in words that would excite as well as edify the public. Carl had to keep explaining that we were not seeing faces and pyramids or other constructed objects on the surface of Mars. Some experimenters were quite jealous of the media attention focused on Carl, but that had become par for the course on planetary encounters.

Viking Impact

Much has been written about the Viking results. The two orbiters mapped 97 percent of the Martian surface with fifty-two thousand images. Lander 2 operated until 12 April 1980 and returned fascinating pictures of water frost forming on the surface during the Martian winter. Lander 1 was still going strong when JPL controllers lost lock with its antenna on 13 November 1983. Between them, the two Landers returned over three million weather reports from Mars over a period of more that seven years, documenting the fascinating cycle of rising and falling surface pressure and water vapor migration from polar cap to polar cap during the Martian year. Almost everything we know about Mars up until 1997 came from the Viking missions. It is Viking's measurement of atmospheric isotopes that let us say with confidence that certain highly interesting (fossil-bearing?) meteorites found in Antarctica originally came from the surface of Mars.

The Viking Project was an enormous success, in spite of the disappointment of not finding life. Everything on both orbiters and landers worked to perfection (except for one of the two seismometers), a tribute to the dedication and determination of people who yearned to produce the first landing on another planet and to make the first in situ search for extraterrestrial life. After a hiatus of more than twenty years, from 1976 to 1997, let us hope that the new generation of American and international Mars missions will carry that quest ever forward.

Viking's final cost—close to $1 billion including launch vehicles—overran the budget initially established for the 1975 missions by 12 percent. In view of double-digit inflation and the multitude of technical advances made I feel no great need to apologize for that. And I take my hat off and salute Jim Martin. He helped make Viking difficult in scope but then pulled it all off with strong, unwavering leadership. (The earlier bold Viking, Leif Eriksson, would have been proud.) I know of no one else, other than perhaps the late George Low, who could have done it. Jim Martin is one helluva strong project manager—and really not a bad guy if you're not bucking against him.

I like to think that the American taxpayers got more than their money's worth out of Viking. In addition to the wealth of terrific data and the increase in our understanding of Mars and the insight it gives us on how Earth-like planets evolve, all those advances in technology that Viking was required to produce have had many broad applications. Remember the hairlike tubing and nearly invisible tiny valves in the biology instrument? The very same valves still made by Parker Hannifin are today being implanted in humans, where they release insulin into the blood stream of diabetes patients on a demand basis without the patient even being aware or having to worry about hypodermic needles and schedules.

The supersensitive tandem combination of a gas chromatograph and a mass spectrometer in the Viking GC-MS instrument has found many applications. While Klaus Biemann, the team leader, was busy at MIT categorizing signatures of various compounds run through the GC-MS in preparation for Viking operations on Mars, a nearby Boston hospital that had read of his Viking development work called him about a young girl patient they had in emergency. She was in a coma and sinking fast, apparently as a result of swallowing some very toxic poison or drug. Their standard lab tests showed nothing. Could Klaus and his

GC-MS help? He would try. A blood sample was rushed over to MIT by motorcycle messenger. In just minutes the GC-MS was able to identify the toxic compound and the girl was treated and recovered. Klaus was swamped with subsequent calls. By 1976 he and his GC-MS were responding to more than fifteen hundred emergencies a year in the Boston area. He had become a local hero. Companies were eager to build commercial units, both fixed and portable, and sell them to hospitals throughout the world. Eventually, other applications included the monitoring of trace pollutants in air and water, keeping tabs on air quality in closed systems (such as those on submarines and manned spacecraft), and the detection of explosive materials in luggage lockers and the cargo bays of airliners.

Viking was expensive, but its scientific returns were enormous and its technology spinoffs many. It earned its place in history.

CHAPTER 7

1977

VOYAGER 1 AND 2 FLYBYS
OF THE OUTER PLANETS

After the ill-fated Voyager Mars Project succumbed in the fall of 1967, I had stepped into my new role as NASA's manager of Advanced Planetary Programs and Technology. In support of planetary program planning my predecessor in Code SL, fellow Notre Damer Pitt Thome, had established a contract with the Illinois Institute of Technology Research Institute (IITRI) to identify possible and promising new planetary missions. The head of the IITRI study group was a tall and impressive gentleman by the name of David Roberts, who was aided by Allan Friedlander and, especially, by young and quietly enthusiastic John Niehoff. Dave and John quickly introduced me to the unique possibilities for outer-planet missions in the late 1970s.

In the 1960s Michael Minovich and Gary Flandro at JPL had explored multiplanet mission possibilities using gravity assist during planet flybys. Flandro determined that the outer planets would be in a rare alignment between 1976 and 1979, which would permit a strong gravity-assist boost from a Jupiter flyby to speed spacecraft to any one of the other outer planets. Going from Jupiter to a Saturn flyby, a spacecraft could either be deflected by Saturn's gravity up out of the ecliptic plane to speed out to Pluto or accelerate on to Uranus and then to Neptune. This latter four-planet Jupiter-Saturn-Uranus-Neptune flyby mission was promptly dubbed the "Grand Tour" by JPL's head of Advanced Studies, Dr. Homer Joe Stewart. Exciting stuff! Just two spacecraft could explore all of the outer planets. The bargain of a lifetime! Actually, many lifetimes, as this alignment of planets occurs only every 176 years—it last occurred while Thomas Jefferson was president. I immediately allocated a good chunk of my APMT funding to more detailed study of the Grand Tour opportunities and technology requirements.

Don Hearth and I quickly agreed that we had a winner here. In fact, we agreed, if we could not sell the Grand Tour mission we should both be fired. We worried that the title "Grand" implied "very expensive," but we could think of no better name. In January 1969 Don and I organized the Outer Planets Working Group (OPWG), composed of the following scientists and engineers from Headquarters and interested NASA centers:

Code SL	Bob Kraemer, chairman
	Don Rea
IITRI	John Niehoff
ARC	Howard Matthews
	Paul Swan
GSFC	Emil Hymowitz
	Jim Trainor
JPL	Jim Long
	Bob Mackin
MSFC	Dan Hale
	Horst Thomae

The group met at least monthly to digest the great quantity of mission-description information generated by the centers and IITRI. Ames concentrated on their Pioneer-class small spinning spacecraft to perform not only flybys but also Jupiter orbiter and atmospheric probe missions. JPL focused on the Grand Tour flybys and recommended an upgrade beyond their Mariner spacecraft technology to provide reliability over mission lifetimes of twelve years and more.

The proposals from MSFC were by far the most ambitious. They assumed that the launch vehicle would be their enormous Saturn 5, topped off with the NERVA nuclear-rocket stage then in early development by NASA and the AEC. To make use of all that payload capacity they proposed launching a cluster of spacecraft that would drop off orbiters at Jupiter and Saturn, probes into Ganymede and Io and Saturn's rings, flyby spacecraft to Uranus and Neptune and Pluto, and even a cluster of probes to Halley's Comet. Quite a flying circus. I was horrified. They wanted to go down the same grand-scale path that had doomed Voyager Mars. I was really not kind to the Marshal people— after all, they had worked hard on their studies. I referred to their proposed mission as either the "Roman Candle" or else the "Grandiose Tour." The OPWG quickly voted it out of contention and focused on the upgraded Mariner-class spacecraft for flyby and orbiter missions and the spinning Pioneers for probe delivery and magnetospheric studies.

In May 1969 the OPWG approved by unanimous vote their conclusions and recommendations. The principal ones were:

1. The unique multiple planet flyby opportunities in the 1976–1980 time period provide an excellent means to survey all the outer planets; these opportunities should be fully utilized. Redundant launches should be seriously considered.
2. The combined J-U-N (Jupiter to Uranus to Neptune) and J-S-P (Jupiter to Saturn to Pluto) Grand Tours are superior to the J-S-U-N Grand Tour. They provide coverage of all five of the known outer planets and they cut the trip time roughly in half (7½ years versus 13 years or more, depending on clearance provided in passing beyond Saturn's rings).
3. A new outer planets spacecraft carrying 100 to 200 lbs. of science instruments appears adequate for accomplishing the more urgent scientific objectives. A test flight of this new spacecraft in 1974 is highly desirable to insure a high state of reliability for the subsequent multiple planet launches.
4. Neither a spin-stabilized nor a three-axis stabilized spacecraft is ideal for *all* planet- oriented and particles and fields (P&F) instruments. A hybrid spacecraft configuration which has both spinning and stable mode capacity appears to have promise from at least the standpoint of science instrument pointing. The concept should be studied and evaluated. (Author's note: this concept was later adopted for the Galileo spacecraft design.)

5. Orbiters and probes of Jupiter should be scheduled as soon as practical. Orbiters of planets beyond Jupiter can, and perhaps should, await flyby results.
6. While atmospheric entry at the major planets is very difficult, preliminary analysis indicates that an entry vehicle with perhaps ⅓ its weight in ablation heat shield material could survive grazing Jupiter posigrade equatorial entry. Only a relatively minor decrease in entry velocity can be achieved with orbital entry rather than direct entry. Much more analysis and data will be required; probe study and technology efforts should be stepped up.
7. The larger satellites of the outer planets may prove as interesting as some of the solar system's planets. (Author's note: Boy, were we ever right on this one. The satellites could very well have all looked like our Moon, but instead they were full of surprises, some truly astounding.)

Selling the Grand Tour

In 1968, after President Johnson had "fired" Jim Webb, former General Electric executive Thomas O. Paine took over for a brief stint as NASA administrator. Code SL set out to get the obviously intelligent and motivated Paine excited about the prospects for exploring the entire solar system. Don Hearth organized a real show for Paine and his senior management council, beginning with a beautiful full-color artist's depiction of the solar system projected across the entire three-screen front of the mini-auditorium while the theme music from the movie 2001: A Space Odyssey played in the background. Don started off our pitch, followed by Don Rea, Bill Brunk, and then me on our proposed flight missions. My presentation on future planetary missions had reached the Grand Tour and was going over very well until I made the mistake of noting that all of the senior managers in the room would be retired before our distant spacecraft reached the planet Pluto. Faces fell. I never made that statement again, realizing that the long time-span of the mission was going to be an even greater demotivator for White House occupants, whose terms lasted eight years at best. In spite of my boo-boo, however, Paine was enthusiastic and we had his support in promoting the Grand Tour to the space science community and the White House.

Our Grand Tour mission description based on the OPWG output was enough to intrigue President Nixon's speech writers and his science advisor. Into Nixon's 7 March 1970 statement on the future space program they wrote, "In the late 1970s the 'Grand Tour' missions will study the mysterious outer planets of the solar system—Jupiter, Saturn, Uranus, Neptune and Pluto. The positions of the planets at that time will give us a unique opportunity to launch missions which can visit several of them on a single flight of over 3 billion miles. Preparations for this program will begin in 1972." Sounded like a presidential endorsement. Still, statements and inspiring speeches by politicians are one thing—getting into an approved budget is quite another. A presidential statement in 1970 in no way guaranteed that our Grand Tour project would make it all the way into an approved appropriations for 1972. We still had to clear the usual budget hurdles, starting with getting a solid scientific endorsement.

With a menu of exciting outer-planet mission possibilities, we now briefed the LPMB and launched a campaign to develop support in the space science community. The Space Science Board of NAS had already performed several studies of outer-planet exploration, including "Planetary Exploration: 1968–1975," reported in June 1968. The SSB now joined with the LPMB in an Outer Planets Summer Study during the week of 10–14 June 1969. Their final report, "The Outer Solar System: A Program for Exploration," as drafted by cochairman Jim Van Allen, endorsed the conclusions of the OPWG but with even more emphasis on atmospheric probes. They also made the very helpful recommendation that our proposed 1974 test flight fly by Jupiter on a gravity-assist trajectory that would take it back over the poles of the Sun (the later joint ESA-NASA mission called Ulysses). We now had formal and solid backing by planetary scientists.

It was time to start heavy marketing with OMB and Congress. Within Code SL, however, we were stretched very thin. Don Hearth departed Headquarters in early 1970 to become deputy director of Goddard, and Don Rea returned as planned to JPL. Glenn Reiff picked this time to transfer to the Department of Transportation, so I found myself filling the roles of director, deputy director (and chief scientist), manager of Advanced Programs and Technology, and program man-

ager for the Helios and Pioneer Projects. My boss, John Naugle, said
he would be very comfortable with me as a replacement for Don
Hearth, but instead he named me deputy director and "acting director"
of Planetary Programs. He was straightforward in saying that he hoped
to attract a top planetary chief scientist and wanted the option of offering him or her the title of director if necessary. While he searched
he let me recruit for the missing engineering managers. I started first
with a manager for Advanced Programs.

In 1961–63, while we at Ford Aeronutronic were busy with the development of the Ranger lunar landing system and three different
Ranger Block 5 lunar payloads, we were also competing with the Northrup Corporation to become the system contractor for the planned series
of Ranger Block 5 spacecraft buses. A young systems engineer from
Northrup, Daniel Herman, emerged as a tough competitor. He was
very sharp technically and showed good marketing sense and tenacity.
He was smooth, but a New York–bred toughness showed through. In
March of 1963 Northrup won the Block 5 bus contract.

Dan was just the kind of person I wanted as SL's manager of Advanced Programs. At a long meeting at his house in Granada Hills,
California, I must have talked fast because Dan and Rhoda Herman
agreed to a cut in income and a move from sunny California to Washington, D.C. Dan did a great job in Code SL, and I am happy to report that he says he never regretted the move to NASA.

Next I managed to talk Fred Kochendorfer (and wife Ruth) away
from Philco Ford in Palo Alto, California, and back to Code SL, where
he had previously served under Oran Nicks as Mariner program manager. Fred was able to pick up immediately for the departed Glenn Reiff.

During this time John Naugle found he had the dubious honor of trying to pick a precious few new space science projects out of a growing
list of very worthy candidates. Any selection he made was bound to stir
a wave of protest from the losers. He wanted some backing for his decisions, so he and Homer Newell decided to ask the SSB for help in setting priorities. (After witnessing the ensuing "war" they had reason to
regret this decision and conclude that they could ask the SSB to evaluate the merits of various missions, but NASA management alone would
have to carry the lonely burden of deciding final funding priorities.)

In the summer of 1970 the SSB did indeed conduct a month-long summer study at Woods Hole on Cape Cod titled "Priorities for Space Research, 1971–1980." Following the death of longtime SSB chairman Harry Hess, NAS named senior scientist Herbert Friedman of the Naval Research Laboratory to be acting chairman of the SSB, and Herb assigned himself to head the summer study. Separate committees were named to make the case for their respective program areas, and an Executive Committee, chaired by Herb, was formed to make the final recommendations on priorities to NASA.

The Planetary Committee, cochaired by Tom Donahue and Mike McElroy, received strong input from Carl Sagan, Bruce Murray, Jim Van Allen, Hannes Alfven, Seymour Hess, John Hall, Tommy Gold, Francis Johnson, Jacques Blamont, Richard Goody, John Findlay, George Pimental, and Von Eshelman. The committee first had to deal with the future of Mars exploration. Viking was headed for big-dollar years, and there were ambitious proposals for future Mars rovers and Mars sample return missions. The committee quickly decided that any big push for future Mars study had to await the Viking results. As to the other terrestrial planets, the MVM '73 mission was underway, so that took care of Venus and Mercury for the immediate future. In spite of a vigorous push by Bruce Murray, members could not get too excited about comets and asteroids, so the Planetary Committee's attention now focused on the outer planets and a lively debate on the Grand Tour versus a concentration on Jupiter with orbiters and atmospheric probes. While the orbiters and probes promised excellent scientific return and needed to be part of any future plan, the committee reached consensus that the rare Grand Tour opportunity was too valuable to miss and should get top priority.

Scientific Tug of War

Now the real battle of the summer study began. The Astronomy Committee eyed the growing budget for Viking and argued that NASA's Space Astronomy Program should get funding at least equal to the Planetary Program. Specifically, Princeton astrophysicist and respected dean of astronomy Lyman Spitzer was pushing for putting an LST, or large space telescope (three-meter mirror or better), into Earth orbit. It was a worthy objective, and I was sure NASA would implement it as soon as technically feasible, but the astronomers were afraid that if the Grand Tour were funded there would be no room in the NASA budget for this LST. They therefore attacked the Grand Tour with great vigor. The war was on.

It was not a pretty thing. In addition to supporting their own project, committee members began to try to shoot down the competing project. My assigned role as a NASA manager at the summer study was just to supply information on candidate planetary missions, not to try to influence the outcome, but I found the developing "attack" atmosphere to be very disturbing. I tried to assure debaters that NASA was clearly going to pursue both of these high-priority missions and that they therefore needed to express strong scientific support for both. If not, they could generate a negative climate that would kill both. Budget cutters in OMB and Congress were always looking for negatives. We were facing scientific suicide.

The Planetary Committee members believed that the Executive Committee membership had been stacked against them, with several strong and persuasive astronomer members. The only planetary representative was Richard Goody, who was concerned that any large project would take all the NASA funding away from smaller scientific missions. This was a reasonable concern, but it did not make Richard a very strong advocate for the Grand Tour missions. Richard suggested that the recommended priorities be tied to NASA's budget levels, and the Executive Committee adopted that approach.

On 15 August 1970 the Executive Committee gave their verbal summary report to Naugle and Newell. Regarding Grand Tour, they recommended that the JPL-class missions be done at higher NASA budget levels, but not at lower funding levels. They recommended deferring the LST pending further development in technology, starting with fabricating a lightweight sixty-inch mirror. The war was over? No way.

The next planned step of the summer study was for the Executive Committee to send out a draft of its summary report to the program committees for review and comment. Then the Executive Committee was to prepare its final report. Instead, Herb Friedman wrote his own draft, and before sending it around for comment and approval as planned, he delivered a copy to the Karth House authorization sub-committee on space, where my friend Frank Hamill notified me of this shortcut. Now, Herb was a respected and honored scientist who served on many high-level and prestigious committees and boards. It was not too surprising that, as a recognized pioneer in X-ray astronomy, he strongly favored Astronomy Programs over Planetary Programs, but in his draft report he wrote that the Grand Tour missions were of rela-tively low scientific value, implying that they mainly served to supply colorful television pictures to fascinate the news media.

This language alone was enough to outrage planetary scientists, and his action of sending his personal draft to Congress did not sit well with his own Executive Committee, the SSB, NAS, and its parent NRC. Per-sonally, I rated it as a definite "dirty trick." Promptly modifying Herb's language, NAS published a report that better reflected the findings of the Executive Committee. It endorsed the Grand Tour, but only at higher funding levels for NASA. This did not satisfy me or many plane-tary scientists.

Then NAS appointed Charles Townes, 1964 Nobel laureate as coin-ventor of the laser, to replace Herb as chairman of the SSB. Townes, who was not an instant enthusiast for the Grand Tour and instead fa-vored a Jupiter orbiter mission, organized a follow-up summer study for 1971 at Woods Hole under the chairmanship of Francis Johnson of the University of Texas. The Johnson committee was directed to per-form a new "Outer Planets Exploration Study" with a special charge to evaluate the Grand Tour proposal. The fourteen-member study team included antagonist Herb Friedman and doubter Charlie Townes, but they were balanced by strong planetary scientists such as Mike Belton, Tom Donahue, Richard Goody, Mike McElroy, Robbie Vogt, and Jerry Wasserburg.

I arranged for presentations to the Johnson committee by JPL and Ames on the outer-planet missions they were studying. While I made no recommendations and tried not to influence the committee conclu-sions, I think the committee members were aware of the OPWG conclusions and knew that I believed the Grand Tour opportunity was too good to pass up and deserved at least a Mariner-class spacecraft. I was shocked to find out later from a committee member that Chuck Sonnet, chief scientist at NASA Ames, had convened a closed hearing with the committee in which he said that I had rigged the presentations and that the Ames Pioneer spacecraft could do everything that JPL proposed with their advanced TOPS (Thermoelectric Outer Planets Spacecraft, an advanced Mariner design) and at a fraction of the cost. I had not torqued the JPL and Ames presentations at all; I thought they gave the committee members a fair assessment of the respective capabilities.

The committee voted to recommend the upgraded Mariner-class Grand Tour as an opportunity too valuable to miss, with Jupiter orbiter and probe missions to follow as soon as they could be worked into the budget for a balanced program of planetary exploration. Their final report concluded that "a program combining TOPS Grand Tour with the more intensive study of Jupiter and Saturn with flybys, probes, and orbiters is fully justified." The final vote on releasing the report was thirteen for and one against, cast by Herb Friedman.

I have long pondered those unthinkable 1970 summer study actions of Herb Friedman. In addition to his outstanding scientific achievements, I have always found Herb to be a cultured gentleman, pleasant to be with, and certainly not an individual one would suspect of rash actions. As acting chairman of the SSB, how could he have even thought of sending his personal version of the SSB summer study to Congress? Certainly as an X-ray astronomer he could be expected to be biased in favor of his own field. Homer Newell in a later report to Jim Fletcher, who was appointed NASA administrator in 1971, complained that as acting chairman of the SSB Herb had too strongly pushed for high-energy and IR astronomy and was generally hostile to large projects. But Herb's maneuver in 1970 went beyond normal bias. It was a dramatic indication of the increasingly fierce no-holds-barred competition for NASA funding among a growing list of scientifically important and attractive science mission candidates.

While the battle of LST versus Grand Tour was raging in the space science community, Ken Weaver authored a marvelous forty-eight-page article in the August 1970 issue of *National Geographic Magazine*. Entitled "Voyage to the Planets," it featured stunningly beautiful and colorful paintings by Ludek Pesek of all the planets and their major satellites. Weaver wrote as an imagined observer riding on spacecraft to the planets, including on our proposed Grand Tour spacecraft. He concluded:

> On March 9, 1986, we encounter Pluto on our Grand Tour. At that time the little planet, probably a snow-covered rock, is thirty times as far from the sun as is Earth, and the solar energy falling on each square mile is a thousandth of that for Earth.
>
> Our long voyage to the planets is ended. The Grand Tour spacecraft moves on to an endless wandering beyond the solar system and into the mazes of the Milky Way. We have traveled farther than any man before us, and have seen such wonders as no eye has ever beheld.
>
> With Immanuel Kant, the eighteenth-century philosopher, we may say: "I have . . . ventured on a dangerous journey, and I already behold the foothills of new lands. Those who have the courage to continue the search will set foot upon them."[1]

Anyone who could read that article without getting excited must be brain dead, I thought. I ordered two hundred copies from the National Geographic Society and sent them all over the United States, including to the OMB staff and all of the members and staff of our congressional committees. (Just educating, not lobbying. All very proper.)

Organizing a Grand Mission

Aided by the 1971 endorsement from the SSB report, by the late summer of 1971 my friends in the White House's Executive Office Building (such as very savvy and helpful Ron Konkel) confirmed that we had OMB strongly sold on the Grand Tour, at least from Deputy Director Caspar Weinberger (later to be appointed secretary of defense) on down—Director Schultz was more obsessed with reducing the total NASA budget. Out in Pasadena, JPL had an excellent team, including Jim Long and Roger Bourke, working on mission design, and Bill Shipley and Ron Draper were preparing the preliminary design for TOPS. The "thermoelectric" in the acronym referred to the use of RTGs upgraded from those that were to power Pioneers 10 and 11 and the Viking Mars Landers. A new feature of the TOPS design was a self-test and repair (STAR) triply-redundant, majority-voting computer that would let the spacecraft diagnose itself and correct any problems that might arise during its decade-long flight.

We settled on two missions, with two launches for each mission. In 1977 we would launch two spacecraft to travel from Jupiter to Uranus to Neptune (J-U-N) and two in 1979 to pass from Jupiter to Saturn to Pluto (J-S-P), reaching all of the outer planets with just two mission sets. One of the beauties of the J-S-P trajectory was that it had to fly by Saturn beneath Saturn's equator in order to be deflected upward out of the ecliptic plane to reach Pluto's tilted orbit. The spacecraft would thereby avoid the potentially hazardous close crossing of the ring plane at Saturn. JPL might have been doing a little lobbying with OMB that they did not tell me about, but in any event the OMB staff liked all aspects of this Grand Tour package and readily accepted it into the fiscal year 1973 budget request President Nixon was to send to Congress in January 1972.

Meanwhile we had been cultivating support in Congress for this major new outer-planets project to follow the Pioneer F/G missions to Jupiter. In my 16 February 1971 testimony to Representative Karth's subcommittee on Space Science and Applications I had laid it on pretty heavy:

> Beyond Jupiter lies the planet Saturn, rivaling Jupiter in size and certainly one of the most beautiful objects in the heavens. Although striking to behold through a telescope, the bright but translucently thin rings around Saturn are a puzzle to astronomers and planetologists. The ten known satellites of Saturn also present challenging questions. For example, the satellite Iapetus varies widely in brightness as it orbits Saturn. It is over six times brighter when it reaches maximum elongation on one side of the planet than when it appears on the other side of the planet. There is currently no plausible explanation for this effect.
>
> Clearly we would like to investigate Saturn as early as it is practical to do so. However, Saturn lies almost a billion miles from the Sun and pre-

sents a challenging mission to currently available launch vehicles. The remaining outer planets, Uranus, Neptune, and Pluto, are still farther from the Sun—up to four billion miles out to Pluto's orbit—so that a direct flight to Pluto would take about thirty years. Fortunately, there is a rare opportunity available to us during the late 1970s to take advantage of gravity assist maneuvers to reach *all* of the outer planets. During the late 1970s the outer planets are so aligned that all of the outer planets can be visited in two successive three-planet flybys. Launch windows in 1976 and 1977 provide an opportunity for a spacecraft to tour from Jupiter to Saturn and on to Pluto, while 1979 is the best opportunity for a Jupiter-Uranus-Neptune tour. These multiple-planet swingbys or "Grand Tours" allow even the farthest outer planets to be reached in nine years or less with Titan-Centaur launch vehicles. These uniquely favorable and economical opportunities to investigate all the outer planets and their major satellites will not become available again for between one and two centuries.

Over the past four years, many types and sizes of spacecraft have been evaluated for accomplishing the exploration of the outer solar system. The Thermoelectric Outer Planets Spacecraft (TOPS) concept for a basic multipurpose spacecraft has been selected as the most economical approach consistent with the highest priority scientific objectives and the long-life reliability required for all outer-planets missions. A self-test and repair (STAR) computer, already well along in development, is capable of adaptively managing and controlling all the TOPS spacecraft functions independent of any commands from Earth. Results to date in the development of the STAR computer and of its highly miniaturized electronic microcircuits already promise an unprecedented advance in operating reliability. This highly compact and competent STAR computer weighing only thirty pounds and "guaranteed" to last at least ten years without servicing is expected to see widespread application not only in many future spacecraft and throughout the aerospace industry, but also in general commercial applications.

We are designing the outer-planets spacecraft to be a multipurpose vehicle which can be readily configured for either flyby or orbiting missions. In its flyby configuration it should also be suitable for carrying deep-entry atmospheric probes to Jupiter during the 1980s. We are continuing to work with the Space Science Board (SSB) and NASA's in-house scientific advisory groups to determine the best sequence of Pioneer and TOPS-class missions consistent with available resources.

NASA currently favors taking advantage of the multiple-planet opportunities with single Jupiter-Saturn-Pluto (J-S-P) launches in 1976 and 1977 and dual Jupiter-Uranus-Neptune (J-U-N) launches in 1979. Recent planetary study groups (NASA in-house, Lunar and Planetary Missions Board, SSB 1969 Summer Study on the Outer Solar System, President's Scientific Advisory Committee, and SSB 1970 Planetary Panel) have supported this general approach.

However, along with this endorsement by the scientific community of the scientific value of the Grand Tour missions, questions have been raised about the appropriate level of funding for the exploration of the outer solar system. During the 1970 summer study sponsored by the Space Science Board on "Priorities in Space Science and Earth Observations," there was considerable debate on the relative priorities that should be assigned to the different fields of space science in a climate of restricted resources. Scientists specializing in planetary sciences and in space physics strongly advocated the Grand Tour missions. Scientists working in other fields understandably campaigned for projects more directly aligned with their own scientific interests. The Executive Committee at that study was faced with the difficult task of deciding relative priorities among a number of such projects, all attractive and promising substantial scientific return. Their conclusion on the Grand Tours, as stated in their final report, was that "these missions are of high scientific merit in a higher level program." Their definition of a higher level program was one representing a substantial increase in the fiscal year 1970 OSSA budget. This conclusion reflects their concern that such a large project could displace valuable smaller projects if the OSSA budget were constrained to the fiscal year 1970 level. As indicated in the president's budget, we have not been constrained to that level. In fact, the fiscal year 1972 budget for NASA provides a 44 percent increase for OSSA compared to the fiscal year 1970 level. At this budget level we believe that we should proceed with a strong program of outer-planets missions.

The extent of exploration and the potential for scientific discovery from the four planned multiple-planet missions is truly outstanding. The spacecraft will not only fly close by all of the five outer planets for detailed observation, but they will also view most of the larger satellites of the outer planets. Some of these satellites are as large as small planets and give some evidence of having atmospheres. We will be able to study many of them much more thoroughly than Mariner 4 viewed Mars during its historic visit in 1965. In the four Grand Tour missions we will achieve at least twenty-seven close encounters of planets and their satellites. By taking advantage of the unique alignment of the planets in the late 1970s we will accomplish these twenty-seven close flybys at the *lowest cost per encounter ever achieved in our Planetary Exploration Program.*

From all of these bodies, the spacecraft will return vital data on surface appearance, size, mass, density, precise position, orientation, temperature, cloud patterns and movements, composition of the upper atmosphere, magnetic fields, radiation belts, and solar and/or galactic wind interactions. The resultant findings will not only represent a major step toward understanding the origin and evolution of the solar system, including our home planet Earth, but should also provide new clues about the origin of the entire universe. As a further help to cosmologists, the outer-planets spacecraft will pass out of the region of space dominated by the solar wind into a purely interstellar medium. Here the true nature of cosmic rays can be studied for the first time outside the influence of the Sun, thereby helping us understand the origin of these extremely high energy particles and their contribution to the nature of the universe.

We plan to hold an industry competition for the selection of a systems contractor to complete the detailed design of the outer-planets spacecraft, to develop it to a high state of reliability, to integrate flight spacecraft and to support flight operations. During fiscal year 1972 we will continue to study outer-planets missions. Both multiple-planet flybys and orbiting missions will be defined in further detail. A careful evaluation will be made to identify further economies that can be realized in the TOPS spacecraft without seriously detracting from its scientific return and potential reliability. Detailed design will begin on the spacecraft engineering model and advanced development will be conducted on long-lead items. The entire effort is scheduled to have the spacecraft developed in time to take advantage of the Jupiter-Saturn-Pluto launch opportunity in 1976. This effort is essential to developing the basic building blocks for the next decade of exploration of the solar system. The work must be started in fiscal year 1972 if we are to take full advantage of the rare and uniquely economical multiple-planet flyby opportunities.

It had always been my intent to have the Grand Tour spacecraft built by a systems contractor rather than in-house at JPL, which was rather busy at the time with Mariners 8, 9, and 10 and with the Viking 1 and 2 Orbiters. I wanted the JPLers lean and hungry and motivated to sell new missions, including a Mariner-class mission to a comet and Mars rovers. One of the contractors who responded with great interest was Boeing. They fought for and won the systems contract on Mariner Venus-Mercury (a very challenging project, both technically and budget-wise) just to impress JPL that they could handle the Grand Tour project. To further encourage Boeing I gave a talk, titled "Solar System

Exploration: A Strategy for the 1970s," at the annual meeting of the American Astronautical Society in June 1971. The meeting was held in Seattle, so I knew many Boeing engineers and managers would be in attendance. I emphasized that

> the president's budget for NASA in fiscal year 1972 includes a new start for Outer Planets Missions, which are currently planned to include Jupiter-Saturn-Pluto Grand Tour missions launched in 1976 and 1977 and a pair of Jupiter-Uranus-Neptune Grand Tours launched in 1979. . . . The Grand Tours constitute probably the most important single contribution to completing the survey portion of a balanced program of any prior set of missions ever proposed by NASA. As currently planned, these four missions can achieve up to twenty-seven close encounters, including all of the outer planets and many of their twenty-nine known satellites. While close encounters with asteroids have not yet been studied in detail, it is believed likely that one or more of these can also be included as a part of the Grand Tour missions.

Then in July 1972 came a surprise from JPL. Bill Pickering, JPL's long-time director, had begun to worry about future work dropping off at the lab. Without warning, he went directly to NASA administrator Jim Fletcher (bypassing me and Naugle) with a request that the Grand Tour project be done in-house at JPL. It was a surprise and a jolt to me—I had been promoting the project to all interested aerospace contractors, not just Boeing, as a promising business opportunity for them. They in response were helping to sell the missions.

To resolve the conflict between me and Pickering, Fletcher appointed his deputy, George Low, as mediator. At our showdown meeting at JPL, Pickering gave his reasons and I presented mine. George, with a long management history at NASA centers, decided in favor of a secure work load for JPL. As a former contractor I felt that NASA had betrayed the aerospace industry in taking away a job targeted for them that they had helped to promote. It reflected on my own personal integrity and I did not much care for Pickering's sudden move without any warning or discussion. Taking the work in-house certainly cost us valuable lobbying by the potential contractors.

I suppose hindsight can be used to support Pickering's position. Missions became much tougher to sell in the shuttle-development era, and

JPL may well have been hurting without the years of very active Voyager effort. On the other hand, under the pressure to keep JPL employed, we might have had a new start for a Comet Halley mission.

Assembling a Headquarters Team

Meanwhile, within NASA Headquarters John Naugle finally gave up on his personal efforts to attract a top planetary scientist for Code SL. The scientists he approached were all deep into active research with secure grants and positions at universities or government laboratories. They saw no compelling reason to join the rat race at NASA Headquarters. Finally, in March 1971, John officially made me director of Planetary Programs and let me do my own recruiting for a chief scientist.

I had serious discussions with respected atmospheric scientist Tom Donahue, then at the University of Pittsburgh, but could not quite entice him to leave his research and university life. Then several planetary scientists suggested that I talk with Ichtiaque Rasool, a planetary atmospheres scientist at NASA's Goddard Institute for Space Science (GISS) in New York City. I had read several of Rasool's papers and popular articles on Venus but had never met him. I especially valued the recommendation from my good friend (and fellow skier and wine taster) Don Rea, who knew firsthand the job of being SL's chief scientist. Off I flew to New York City.

The situation at GISS proved most interesting. The director there was the renowned scientist and author (of, for example, the very popular book on astronomy *Red Giants—White Dwarfs*) Bob Jastrow. I quickly got the impression that Bob really had Rasool pretty much running the place, freeing himself to pursue his own professional interests. He was not a bit happy that I had come to talk to his right-hand man. But he let me. I found Rasool not only easy to talk to but also receptive to the prospect of progressing within NASA beyond the confines of Jastrow's GISS. He discussed it with his lovely French wife Françoise and then agreed to the move.

Ichtiaque Rasool is without doubt one of the most colorful characters in all of NASA. One could write a book just about him. I am not going to try. Born in what is now Pakistan, he worked in Karachi as a meteorologist, trying to predict monsoons. Moving to France, he earned a doctorate degree in atmospheric science from the University of Paris. Then to GISS. It is hard for me to imagine him handling the administrative chores for Bob Jastrow. The freewheeling Ichtiaque shuffling papers? Hard to picture. For example, soon after he arrived at NASA Headquarters we received a notice from New York City requesting to garnishee his pay so they could recover fines from several hundred parking tickets he had ignored. Ichtiaque had just considered them a bureaucratic nuisance and thrown them away, just as he had in Paris. We suggested that it would be more dignified if he would settle with New York. He did.

In spite of this indifference to paperwork, and the distraction of his instant appeal to most of the women he encountered, Ichtiaque did a great job as chief scientist for Planetary Programs. He was solid, well respected technically, and well liked throughout the planetary science community. His affability was all-important when he had to soothe a frustrated scientist whose instrument proposal had been rejected. Although Ichtiaque, Dan Herman, and I were quite different from one another, we got along very well and were, I believe, an effective team in helping to create a strong planetary program. During the early 1960s the SL team of Nicks-Lidell-Hearth was very productive, but Urner Lidell was not all that effective in relations with planetary scientists. In the later 1960s the team of Hearth-Rea-Kraemer was, in my not-so-humble opinion, outstanding. In 1970 the one-man team of Kraemer-Kraemer-Kraemer was obviously overworked if not overstressed, but by 1971 the new SL leadership of Kraemer-Rasool-Herman was ready to charge into the challenging and exciting 1970s.

For the position of Grand Tour program manager I had an eager volunteer in Warren Keller, one of the very best program managers from NASA's Office of Aeronautics and Space Technology (OAST). Warren himself selected a dedicated engineer from OAST, Art Reetz, as his deputy. I in turn assigned Milton "Mike" Mitz, one of our very best Code SL scientists, to serve as Grand Tour program scientist. We now had a solid Headquarters team to work in partnership with the growing Grand Tour crew at JPL and were really rolling.

A Major Setback

Confident that we had sold the Grand Tour, I scheduled myself for back surgery on 1 December 1971. When I was twenty-two years old I had damaged one of my lumbar vertebrae. After a year of recovery, and thereafter occasionally wearing a back brace, I had lived with it through twenty-one years of skiing, body surfing, and sailing (I even managed to carry my bride across the threshold in 1954), but now my entire left leg was growing numb. I was told by three different orthopedic experts that I either had to have three vertebrae fused together *now* or else lose all use of my left leg. So I was relieved that I could take some time off now to get the surgery performed. The fusion procedure by orthopedic specialist Dr. Leo Van Herpe went well, and two days later I was past the morphine stage and resting comfortably in my hospital bed at Georgetown University Hospital. The bedside phone rang. It was Ichtiaque Rasool, who was acting director in my absence. He said, "Bob, you won't believe it, but they have canceled the Grand Tour."

Poor Ichtiaque, he felt he was responsible. In fact, he was not even in the decision loop. What had happened was that NASA was proposing in the new budget to start ramping up on funding to develop a reusable space shuttle. Deciding that they could not push the NASA budget up that far in the coming year, OMB imposed a cut in NASA's budget request, anticipating that NASA would slow down the starting effort on the shuttle. Jim Fletcher was heavily influenced in budget matters by his very capable comptroller, Bill Lilly, who was convinced that NASA's public support rested primarily on manned spaceflight. That wise adviser to several NASA administrators, Willis Shapley, shared the same opinion. In response to their advice Fletcher made the decision to meet the new OMB budget mark by deleting the Grand Tour rather than slowing the shuttle. (I have no evidence that Fletcher intended a "Washington Monument bluff," named for the maneuver practiced by the Department of Interior, which when faced with a budget cut, always responds that it will have to close down the popular Washington Monument.)

If we were stunned in SL, so were the people at JPL. Roger Bourke recalls his feelings: "The most unpleasant surprise was the lack of support in Washington, not necessarily by our immediate contacts in the program office. As 'kids' in our 20s or so, it was inconceivable that there would be anybody who wouldn't fall in love with this idea."[2]

The people at OMB were equally shocked. They were totally sold on the Grand Tour Project and had not even imagined that NASA would cut it. They made the unprecedented move of offering to *restore* funding in the NASA budget if we could come up with a somewhat reduced cost for the Grand Tour. Rasool and Herman got to work on scaling down the project with the Grand Tour team at JPL, and I joined them shortly. The STAR computer had to go, and with that we felt we could not guarantee that the spacecraft would last more than the four years they would take to reach Saturn. We were back to an evolved Mariner spacecraft. We regretfully dropped the Grand Tour name and relabeled the mission Mariner Jupiter-Saturn (MJS), with two identical spacecraft to be launched in 1977. Highly respected university scientists Jim Van Allen and Mike McElroy teamed up to give an excellent presentation on the mission's scientific potential to the SSB, and on 22 February 1972 we had the vital SSB endorsement. This was followed immediately by endorsements from the President's Scientific Advisory Committee (PSAC) and the White House Office of Science and Technology (OST). Congress readily approved, and we had a project. Fletcher and his staff somewhat later decided to jazz up the project name (Mariner sounded old hat) and resurrected the name Voyager to replace MJS.

This sounds like we had given up on a Grand Tour. I certainly had not, and neither had the engineers at JPL. We still had two spacecraft in the project, and there was no way of knowing how long they might continue to function in space. A J-S-P flight and a J-S-U-N flight were still possible, but we were not going to guarantee it or even publicly promote the possibility. If at least one of the Voyager spacecraft reached Saturn the project would be officially declared a success, which was important to the record of both NASA and JPL. In my mind, though, I remembered Don Hearth and I agreeing that we both should be fired if we did not succeed in selling a Grand Tour.

Over the next two years I made several presentations to Fletcher trying to sell him on adding a Mariner Jupiter-Uranus launch (MJU 79), but to no avail. His stated objection was that the trip time was too long to gamble on, but a more probable reason is that he did not want to buy another Titan-Centaur while he was trying to shift all NASA launches to the new shuttle.

Getting on with Voyager

In Code SL at Headquarters we had manned up the strong Voyager team of Warren Keller, Art Reetz, and Mike Mitz. All very good people. Warren Keller, for example, later went on to become the very popular director of NASA's important Wallops facility near Chincoteague, Virginia. At JPL Bob Parks wanted to appoint the up-and-coming John Casani as Voyager project manager. John had done an impressive job as Spacecraft Systems manager on Mariner 10 and was clearly project manager material. Nevertheless, I objected. I had no doubts about Casani's abilities, but I knew we were going to have some tough times ahead with George Low, and I wanted to at least start the project with an experienced project manager who could cite his own past successful experiences. No one could be better than Bud Schurmeier. Parks wanted to name Schurmeier to be his deputy for all flight projects at JPL, but instead agreed to my request and appointed Bud Voyager project manager.

Sure enough, the brilliant and oh-so-tough George Low went after the new Voyager Project as a place to highlight his drive to lower the cost of space projects. George correctly forecast that Voyager was going to have to continue to compete with the shuttle within a tight NASA budget. We had already reduced the scope of the spacecrafts' capabilities and did not want to eliminate any chance of them living past Saturn. And the budget had no fat in it beyond a reasonable contingency fund for the project manager. I had held back no Headquarters reserve. George in effect lined Bud and me up against the wall and started shooting verbal bullets. I could not have stood alone against that barrage, but Bud was firm in defending every feature of the design and quoting figures from his past Ranger and Mariner projects to support the need for at least a 10 percent budget reserve. George reluctantly backed down. We now had a firm plan, a realistic budget, and full go ahead from NASA top management.

Pinning down the science scope and payload was the next vital step. We needed a strong and technically versatile scientist to fill the key role of project scientist. Edward Stone, a physicist at Caltech, was highly recommended, but he was a full-time campus professor, not a member of the staff at JPL. It took a lot of persuasion, especially by Bud Schurmeier and Ed's good friend and associate Robbie Vogt, but we finally talked Ed into taking on the job, dividing his time between JPL and the Caltech campus. I blush a little when I recall assuring him that the Voyager job would only take about 10 percent of his time. Boy, was I off on that one. Fortunately, Ed was more realistic in assuming the project would take 30 percent of his time. Voyager eventually consumed Ed, taking 100 percent of his attention during encounter periods, but I do not believe he would trade the experience for anything. In fact, he calls the experience "a journey of a lifetime." He did a fantastic job with both the science teams and in conveying the Voyager revelations to the news media and directly to the public. Following Voyager, he directed the design and building of the spectacular Keck telescopes on Mauna Kea and is continuing his outstanding contributions to planetary exploration as director of the JPL he served so well during Voyager.

Bud Schurmeier assembled an outstanding Voyager design, development, and operations team at JPL with Ray Heacock in the key technical role as Spacecraft Systems manager, TOPS-technology leader Bill Shipley as his deputy for Development, and Ron Draper as Spacecraft Systems engineer. Jim Long, who had so effectively led the early mission studies, continued as Science manager. (Sadly, cancer soon took the life of Jim, who had contributed so much to the creation of Voyager.) The multitalented Chuck Kohlhase directed mission analysis and design, building on the earlier work performed by Roger Bourke, and Dick Laeser managed mission operations. Others in key roles included Ek Davis, Jim Scott, Mike Sander, Bill Fawcett, and Mike Devirian. A very strong team indeed—Bud had chosen well. With the project rolling smoothly, on schedule and within budget, I had no objection at all in

April 1976 when Bruce Murray, newly appointed director of JPL, named John Casani Voyager project manager, replacing Schurmeier, whom he had asked to head up JPL's new endeavors in civil systems (meaning primarily government-funded energy R&D).

Competition among scientists for inclusion in the Voyager mission was fierce, as might be expected. Following the standing policy of judging strictly on scientific merit and readiness for flight, not on reputation or politics, Ichtiaque Rasool and Mike Mitz worked with JPL and the SSSC chaired by Naugle's science deputy, Henry Smith, to select the best proposals:

Science Teams	*Principal Investigators and Team Leaders*
Imaging Science	Brad Smith
Radio Science	Von Eshelman
Plasma Wave	Fred Scarf
Infrared Spectroscopy and Radiometry (IRIS)	Rudy Hanel
Ultraviolet Spectroscopy (UVS)	Lyle Broadfoot
Photopolarimetry (PPS)	C. F. Lillie
Planetary Radio Astronomy	Jim Warwick
Magnetic Fields	Norm Ness
Plasma Science	Herb Bridge
Low-Energy Charged Particles	Tom Krimigis
Cosmic Ray	Robbie Vogt

A name conspicuously and painfully absent from this list was that of James Van Allen. No one had campaigned longer and harder for outer-planet missions and for Voyager in particular, but his proposal just had not scored all that well, and we were selecting on technical merit only. The best we could do was to invite Jim to participate as an interdisciplinary scientist.

Notable in the list of PIs is that the always difficult and demanding Norm Ness had once again beat out his persistent rival, Ed Smith, for the magnetometer instrument. Later, during the development of the spacecraft, John Casani, who was always looking for excuses to have team-building parties and events, organized a little contest. On a large poster board, an artist had drawn a huge and fierce screaming eagle swooping down on a tiny mouse. The mouse was shaking his raised fist at the eagle, with his middle finger raised in a rather vulgar gesture of defiance. The mouse was saying something to the eagle, and the contest was to come up with the mouse's words. The winner of the contest identified the eagle as Norm Ness and the mouse as John Casani, with the mouse saying, "No, Norm Ness, you may not have a forty-meter boom for your damn magnetometer!" Norm was not unhappy over the accurate depiction, and John Casani felt the response to the contest was evidence of good morale and a growing excitement at the anticipated discoveries of the Voyager missions.

A Telescope Offspring

I would like to digress at this point to a related side story—about a telescope. Planetary astronomers always have had to struggle for viewing time on major telescopes. To stellar astronomers, our Sun is a pretty ordinary star. Their priority is trying to understand the origin of the universe rather than the origin of rocky bodies and gas balls around this one mediocre star. To get more planetary observing time NASA, under the leadership of Roger Moore, Ronald Schorn, and William Brunk, had over the years helped to build and operate several telescopes largely devoted to planetary viewing. These included an excellent 61-inch telescope in Arizona built by Gerard Kuiper, the undisputed leader in planetary astronomy, and the 107-inch McDonald telescope in western Texas. Starting in the late 1960s the SSB had been recommending that NASA build a large telescope designed specifically for planetary observations in the infrared wavelengths. A first-class facility with an 88-inch mirror would cost no more than a relatively modest $6–10 million. Each year I would put the proposed Infrared Telescope Facility (IRTF) into my SL budget request, but each year it would get postponed in the final budget squeeze. I needed to find some anchor to keep it from slipping.

During the final debate on the NASA budget request in December

1974 (on a weekend, as usual) John Naugle said he would give me thirty minutes more to come up with a convincing argument why IRTF could not wait yet another year. On an instant inspiration, I called SL's chief of Planetary Astronomy, Bill Brunk, at his apartment and asked him if there was a good Jupiter viewing period in 1977 before the launch dates for Voyager. He responded in the affirmative. I rushed back to Naugle and reminded him that as the two Voyager spacecraft approached Jupiter in 1979 we could target each spacecraft to pass close by one of the four large Galilean satellites. But we did not know which one might have the greatest scientific interest. If we proceeded with building the IRTF in the next year we could make important new observations of those satellites in time to target the Voyager launch trajectories to the most interesting satellites. Better science. John liked that, and the IRTF was in the budget at last.

I will not go into the details of all the difficulties we had with the IRTF Project. With lots of expert help and guidance from Gerard Kuiper we performed careful tests on mountains all over the world and confirmed that fourteen-thousand-foot Mauna Kea on the big island of Hawaii was clearly superior for telescope viewing, especially in the infrared, where avoiding water absorption was so important. (It surprised me that a mountain surrounded by water could be so outstandingly dry above fourteen thousand feet.) We had to fight off a heavy political attack from senators from California and Arizona who wanted the telescope built in their home states. Then the congressional committees tried to take the project away from Code SL and put it under the management of NASA's Construction of Facilities people—the committees said it was not an "instrument" but a "facility," which they liked because they got to name facilities after honored congressmen. The facility people at NASA were great on building steel and concrete structures, but they did not know beans about telescopes. Then the stellar astronomers got into the act and pressured us to upgrade the design for stellar work and guarantee them some time on the telescope. We finally gave in to the upgrades, but that pushed the cost up close to our absolute ceiling of $10 million. Meanwhile, we had contracted the design and construction job through the University of Hawaii, but they were bogging down until I convinced Jerry Smith, an excellent optics engineer from JPL, to take an assignment to the university as their IRTF project manager. Things were finally rolling well when Tony Calio was appointed OSSA deputy and tried to demonstrate that he was a "tough manager" by canceling IRTF as a project in trouble. Well, I am happy to say that the IRTF survived Tony's attack and was built, that it was and probably still is the best IR telescope in the world, and that it has produced important new results on the planets, satellites, and stellar objects and in 1997 identified important new chemical constituents in the brilliant Comet Hale-Bopp. All of this thanks to a timely boost from the Voyager Project. So although the IRTF was finally sold on the basis of contributing to the scientific return from Voyager, it became operational too late for that particular role, and it finally was Voyager helping IRTF rather than vice versa.

Hello Out There

As the Voyager spacecraft were being assembled at JPL, Carl Sagan suggested another greeting to any extraterrestrial finder. The plaques that he, Frank Drake, and Linda Salzman had prepared for Pioneers 10 and 11 had been enormously popular, so Carl's new proposal for a gold-plated phonograph record on each Voyager spacecraft was readily accepted. The twelve-inch records included salutations in almost sixty human languages and one whale language, and much more. For the full story, read Carl's *Murmurs of Earth*. William Burrows provides a summary: "There were thirty-eight sounds, from rain to Morse code to auto gears to a kiss; 115 images, including a fertilized ovum, a snowflake, Bushman hunters, rush-hour traffic in India, DNA structure, and a supermarket; and ninety minutes of eclectic music, including selections from The Magic Flute, the Second Brandenberg Concerto, 'Johnny B. Goode' by Chuck Berry, Senegalese percussion, a Navaho night chant, and a Peruvian woman's wedding song. The images and sounds were selected by a committee headed by Carl. Each record came with an aluminum jacket, cartridge, needle, and instructions."[3]

Carl died on 20 December 1996, much too early in his remarkably productive life. At a lovely and moving memorial service for Carl on

27 February 1997 in New York's awesome Cathedral of St. John the Divine, it was most fitting that the orchestra played several of the classical selections from Voyager's phonograph record.

Humans versus Robots

Two Voyager flight spacecraft were delivered to Cape Canaveral on schedule in the summer of 1977 and checked out for launch. It has always been JPL's practice to perform a great deal of spacecraft reassembly and checkout at the cape. Other NASA centers, such as Goddard, dispute this practice—they say do not fiddle with an already tested spacecraft—but who can argue with JPL's successes? Besides, some teardown was necessary due to vapors from spray painting outside the assembly area that seeped into the clean room and contaminated a number of sensitive detectors in the science instruments. So there was quite a bit of hardware swapping before launch, but the two craft were ready on schedule. The first spacecraft to be launched would be the second to reach Jupiter, so NASA decided to avoid future confusion by calling it Voyager 2 rather than Voyager 1. Voyager 2 was sped on its path to Jupiter atop a powerful Titan 3E–Centaur on 20 August 1977, and Voyager 1 followed on 5 September. With the two Voyagers on their way, JPL advanced their Voyager spacecraft expert Ray Heacock to become Voyager project manager, freeing John Casani to become the project manager for the new and very ambitious Galileo Jupiter Orbiter and Probe Project.

Now began a saga of humans versus the robot machines they had created. Although JPL had not been able to afford the development cost of the originally proposed STAR computer, they had designed a great deal of artificial intelligence into Voyager's brain, called the Computer Command Subsystem (CCS). Even with radio signals traveling through space at the speed of light it was going to take on the order of three hours for round trip communications with the spacecraft at Saturn distance, and we would be facing more than eight hours if Voyager made it to Neptune. A spacecraft could die before controllers at JPL even knew there was a problem. So the CCS was designed to sense any prob-

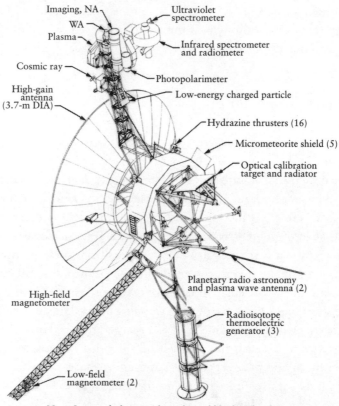

Imaging, NA

WA

Plasma

Cosmic ray

High-gain
antenna
(3.7-m DIA)

Ultraviolet
spectrometer

Infrared spectrometer
and radiometer

Photopolarimeter

Low-energy charged particle

Hydrazine thrusters (16)

Micrometeorite shield (5)

Optical calibration
target and radiator

High-field
magnetometer

Low-field
magnetometer (2)

Planetary radio astronomy
and plasma wave antenna (2)

Radioisotope
thermoelectric
generator (3)

Note: Spacecraft shown without thermal blankets for clarity

The Voyager spacecraft, powered by three RTGs and carrying
a versatile array of scientific instruments for exploring a
variety of outer planets and their numerous satellites.
Launched in 1977, Voyager 1 explored Jupiter and Saturn;
Voyager 2, in a journey hailed as the greatest voyage of
discovery of all time, completed a Grand Tour of Jupiter,
Saturn, Uranus, and Neptune. (Courtesy NASA)

lem with the Voyager subsystems and react quickly to protect the
spacecraft.

Soon after launch our spacecraft started turning itself off. There were
no apparent problems with subsystems, so JPL reactivated the spacecraft, only to have the CCS shut everything down again. Like Dr.

Frankenstein, had JPL's engineers created a monster with a destructive mind of its own?

Faced with a potential catastrophe, a worried Bruce Murray turned like his predecessor to JPL's most experienced project manager, Bob Parks, to take over the Voyager Project. Bob picked two deputies to work together as a team, Ray Heacock, who had guided the spacecraft design, and Pete Lyman, who had done such an outstanding job with mission operations on Viking. They turned to the creators of the CCS, such as Ted Kopf and Chris Jones, who determined that the CCS had just been made too sensitive to shifts in operating parameters. It was interpreting even normal small changes in temperature and power, or even the sighting of tiny sunlit dust particles, as anomalies and was "safing" the spacecraft. Moreover, it was proving difficult to talk to what Murray had come to call their "semi-intelligent and maddeningly literal-minded robot." A skilled team was assembled to work overtime writing new software to upload to the CCS.

The Parks-Heacock-Lyman project team was not to get much relief from pressure. On 5 April 1978 a routine, scheduled ground command to Voyager 2 was inadvertently omitted, causing the CCS to assume that its primary radio receiver had failed, so it switched from the primary receiver to the backup receiver. Unfortunately, the backup receiver refused to lock onto the uplink signals from Earth. Bad news. Voyager 2 was now on its own with no input from its human parents. After seven days the CCS, as preprogrammed, tried and succeeded in switching back to the primary receiver, but for unknown reasons thirty minutes later Receiver 1 failed, never to function again. Voyager 2 switched itself back onto Receiver 2, which could not lock onto signals from Earth. JPL engineers were able to determine that a small tracking loop capacitor had failed in Receiver 2, so that it could only lock onto a very narrow 96 Hertz spread in frequency, and that frequency window would vary with time as the receiver temperature changed. They were able to find that narrow window and get an uplink back to the spacecraft. They then worked up an elaborate computer program that would predict how the receiver temperature (and the receiving window) would vary as the spacecraft temperature changed during flight away from the Sun, including the varying Sun-spacecraft orientation, and as various instruments and subsystems were powered off and on. Their prediction had to be accurate to .1 degree Celsius. Moreover, they had to vary the uplink frequency to correct for Doppler shift due to the changing velocity of the spacecraft relative to the DSN receiving stations on Earth. (It is Doppler shift that causes a train whistle to vary in frequency as the train approaches and recedes.) This fantastic piece of engineering effort worked with amazing accuracy and ensured normal communications all the way to Neptune and beyond. The partnership of humans and their robot creations was working again.

Jupiter Spectaculars

A number of beautifully illustrated books have been published about the discoveries of the Voyager missions (Carl Sagan's *Pale Blue Dot* is one of my favorites), and I will not attempt to repeat what they have covered so well. Let me just touch briefly on Io, the innermost of the Galilean satellites of Jupiter. As Voyager 1 approached Io it took what was quickly named the "pizza picture." With a standard amount of photo-lab enhancement, Io appeared startlingly bright in shades of yellow, red, orange, and white accented with small black markings. Like nothing ever seen before (or since). Imaging Team leader Brad Smith essentially named the picture when he observed that it looked "better than a lot of pizzas I've seen." My astronomer friend Dave Morrison labeled the bright red spot in the center of the image the "piece of pepperoni." Then a young member of the navigation team, Linda Morabito, noticed a strange "bump" along the smooth curve of the rim of Io. On closer examination it was a fountainlike spray pattern, like from a lawn sprinkler. She had found an active volcano on Io! Astounding. Further examination of images revealed at least eight active volcanoes. Morrison's "pepperoni" was one of the largest and was named Pele, after the Hawaiian goddess of volcanoes.

This previously unheralded medium-sized satellite was revealed as one of the wildest bodies in the solar system, with volcanoes spewing colorful sulfur and sulfur dioxide all over its surface. The excitement at JPL was electric, at least equaling the moment when Viking 1 made the first landing on Mars. (Perhaps the first landing on Mars was a bigger event for us engineers, but for our planetary scientists the discovery of active volcanism so far from the Sun just blew their minds.) You might have expected such excitement in the always enthusiastic Lonnie Lane, but even the very stable Ed Stone and Brad Smith were talking in superlatives. The usually poker-faced man-of-few-words geologist Larry Soderblum enthused, "This is incredible. The element of surprise is coming up in every one of these frames. I knew it would be wild from what we saw on approach, but to anticipate anything like this would have required some heavenly perspective.[4] I think this is incredible."[5]

With an impressive data rate of 115,200 bits per second the Voyagers were able to return eighteen thousand pictures of Jupiter's mini–solar system, and the discoveries kept piling in. The Galilean satellites were surprisingly individual in their characteristics. Ultrasmooth Europa had a skin of water ice that appeared to have been cracked by upwelling water surging just below the frozen crust, leading to speculation of a warm ocean and even living organisms beneath the ice. (Arthur C. Clarke promptly wrote two sequels to his classic *2001* based on this possibility.) Io reacted with Jupiter's magnetosphere to create a flux tube carrying an enormous current of one million amperes. Jupiter's colorful clouds swirled with great velocity and turbulence, producing images worthy of hanging in art galleries. The Giant Red Spot in the clouds was a rapidly rotating hurricane large enough to swallow three Earths. Jupiter was found to have rings, joining the family with the other gaseous planets of the outer solar system.

On to Saturn

Bob Parks handed the project management reins back to a confident Ray Heacock, and Ray carefully nursed the Voyagers to Saturn in good shape, Voyager 1 arriving in November 1980. The clouds of Saturn were not as interesting as Jupiter's, although they blew at up to 1,770 kilometers (1,100 miles) an hour, but the rings were certainly anything but disappointing. Not only were there hundreds of sparkling ringlets made up mostly of water ice particles but also some of the ringlets braided and unbraided themselves from day to day. (Maybe Saturn is a female planet braiding her tresses and not the fierce cannibalistic male god of mythology.) Once again, just as at Jupiter, the satellites were at least as interesting as the planet, and certainly more surprising. For example, Mimas was seen to be relatively smooth except for one huge impact crater that covered much of one hemisphere. It was quickly likened in appearance to the Empire's Death Star in George Lucas's *Star Wars* movie series.

A prime objective for Voyager 1 was to explore Titan, the largest moon of Saturn and comparable in size to the planet Mercury. If Voyager 1 failed, the trailing Voyager 2 would be targeted to Titan, ruling out any further Grand Tour. Happily, though, Voyager 1 performed as planned. Titan was confirmed to have a dense atmosphere, more than half again thicker than Earth's, composed of mostly methane and nitrogen, much like the early atmosphere of Earth. A likely prediction was for methane rain to be falling on an ocean of liquid ethane covering a seabed of organic tar. Titan was going to acquire priority as a promising place to study the chemical evolution conducive to life. (The Cassini spacecraft's Huygens probe is on its way to land on Titan in November 2004.)

Voyager 2, now under the management of JPL's Ek Davis, passed Saturn in August 1981. Having completed its basic Jupiter-Saturn mission, the Voyager Project was now officially a "success." That would be an understatement, however. Even the ordinarily unemotional magazine *Science* enthused, "For sheer intellectual fun there has never been anything quite like the Voyager encounters. Volcanos on Io, ringlets around Saturn, *braided* rings—the observations are outrageous."[6] After its triumphal passage of Saturn the Voyager 2 spacecraft was ready to head on to Uranus and Neptune. Could even a cold-hearted OMB turn down such an historic opportunity?

Yes they could. David Stockman, with the incomprehensible support of Dr. George Keyworth, the president's Science Advisor, made a move to not only cut off Voyager funding but also shut down the Deep Space Network and end all of NASA's space exploration. Wearing its Caltech hat, JPL encouraged support from sympathetic congressmen, and NASA's director of Code SL at that time, Geoffrey Briggs, alerted the space science community, including NASA's Physical Sciences Committee (PSC), of this threatening attack. The resulting uproar of protest was impressive indeed. In December 1981 OMB added more than $80 million to the NASA budget to continue operating the Voyagers.

Voyager 2 sailed on into deeper space, but the pursuit of exploration is never serene. Just 102 minutes after the craft's closest approach to Saturn the Voyager 2 scan platform jammed. This is the compact platform that had pointed the cameras to hundreds of positions for the Saturn encounter. Without that pointing capability, imaging science at Uranus and Neptune would be decimated, as would pointing of the other instruments (PPS, IRIS, and UVS) mounted on the platform. Earlier there had been sticking of the scan platform on Voyager 1 and JPL engineers had started lab testing on an identical platform and its associated actuating electric motors and gear train. After being put through the same cycles that Voyager 2's platform had experienced, the actuator mechanism failed due to a breakdown in the lubricant in the gear train. The engineers were able to unjam the actuator in the lab by subjecting it to alternate cycles of heating and cooling. They sent commands to Voyager 2 to try the same technique in space. It worked. The platform was moving again. Techniques were worked out to avoid any incipient jamming in the future, and scanning at Uranus and Neptune was picture perfect. Humans and their robots were working it out together once more.

Next Stop Uranus

Dick Laeser, who had been mission director, now got his turn as Voyager project manager and directed Voyager 2 to Uranus while Voyager 1 was speeding out of the ecliptic plane and out of the solar system. Much had to be done to prepare Voyager 2 for Uranus. Remember, it had only been designed to reach Saturn. Getting images back from Uranus was going to be a challenge. To help the data rate, two of the DSN twenty-six-meter antennas were enlarged to thirty-four meters

Miranda, a satellite of Uranus, as viewed by Voyager 2. Miranda appears to be the most tortured body in the solar system, with deep rifts and abrupt sharp angles unlike anything seen before, including a sheer cliff into a canyon more than nineteen kilometers (twelve miles) deep, ten times the depth of Earth's Grand Canyon. (Courtesy NASA)

and the Australians volunteered to link their sixty-four-meter Parkes radio telescope to the DSN antennas in Canberra. Ways of compressing and preprocessing the image data were perfected, and errors in the faint signal were eliminated by using the latest technology of Reed-Solomon coding. Camera exposures were going to have to be longer in the dimmer light (at 20 AU, only one-four-hundredth the intensity at Earth), so JPL engineers devised ways to control the normal slow swinging of the spacecraft to achieve image motion compensation and eliminate any blurring of the images. It all worked, and the pictures from Uranus were beautifully sharp.

Uranus is unique among all of the planets in that it is lying on its side with its north pole currently pointing toward the Sun. Solar heating on the north pole and a very cold southern hemisphere would be expected to create an unusual weather pattern. The lack of distinctive clouds on a rather bland greenish blue Uranus was a bit disappointing, but other aspects of Uranus were not, such as a magnetic dipole field tilted a surprising sixty degrees from the axis of rotation. In addition, two new rings were discovered along with ten new moons.

One of the moons, Miranda, was the hit of the show. Discovered in 1948 by astronomer Gerard Kuiper, it is only about 480 kilometers (300 miles) in diameter, but its surface as revealed by Voyager almost defies description—a violently tortured mixture of deep rifts and sharp angles unlike anything ever seen before. One shear-walled canyon cleaving the surface was 19 kilometers (12 miles) deep, over ten times the depth of Arizona's Grand Canyon. Geologist Larry Soderblum was astounded once again and proclaimed, "If you took all the bizarre features in the solar system and put them on one object, that would be Miranda."[7] Larry calculated that if a man fell off that 19-kilometer cliff, with Miranda's lower gravity it would take nine minutes before he hit the bottom of the canyon. A strange world indeed. Once again a satellite had overshadowed its parent planet.

After Voyager 2's closest encounter with Uranus on 24 January 1986 I departed JPL and took some vacation time to drive with my wife Anne up the coast of California to the lovely town of Arroyo Grande to visit an old friend of the family. At 8:38 A.M. PDT on Tuesday 28 January I was shaving while Anne was watching the televised launch countdown for the shuttle Challenger at the cape. Anne called out, "You had better come watch, something is happening." One look at the screen and I knew we were witnessing a disaster.

Just an hour and a half later, at 10:00 A.M., the big wrap-up news conference was scheduled for JPL's Von Karman Auditorium to summarize all of the great science findings of the Voyager Uranus encounter. It has been written that after the initial horror at the Challenger loss of lives, the general reaction at JPL was bitterness that shuttle had stolen the news away from Voyager. Now that may possibly have been the reaction of some of the scientists who would have made front-page news with their results, but I believe the management at JPL shared with me another deeper concern. While still watching the repeated television replays of the Challenger explosion, I thought, Goodbye Galileo launch. The 1982 Galileo Jupiter Orbiter-Probe Project was the last new mission we succeeded in getting into the NASA budget before I left NASA Headquarters. It was scheduled for the very next shuttle launch and was now certain to be facing extended delays and major design perturbations. I'll summarize that grueling history in the epilogue.

All the Way to Neptune

Following Voyager 2's Uranus encounter, Norm Haynes, the youthful-looking veteran of numerous JPL missions, took over as Voyager project manager. Voyager 2 was now headed for remote Neptune. The spacecraft was in surprisingly good shape, as long as that one remaining radio receiver kept perking. But the radio transmitter was going to need help to get good images back from Neptune at 30 AU. Attention was focused on increasing the gain of the Earth receiving stations. The three sixty-four-meter antennas of the DSN were given skirts to increase their diameter to seventy meters. Then the twenty-seven antennas of the Very Large Array in New Mexico, along with other antennas around the world (including Spain, Japan, and Australia), were functionally ganged together to form the effective receiving aperture of one very large antenna. Marvelous engineering.

In August 1989, after a spectacular and suspenseful twelve-year journey of 7.2 billion kilometers (4.5 billion miles), Voyager 2 was finally approaching the last planet on its Grand Tour, Neptune. The spacecraft engineers by now had the motion of the spacecraft down to one-sixtieth the rate of movement of the hour hand on a clock. They could nod the cameras to eliminate motion blurring, even during the required long exposure times in the dim light (only one-nine-hundredth as bright as at Earth). All of this effort permitted Voyager 2 to return clear sharp images, revealing that the blue planet had much more interesting cloud structure than Uranus, including large circulating dark and light spots to compare to the Giant Red Spot on Jupiter. Neptune's rings were found to exist in segmented arcs only thinly connected as complete circles.

Once again the biggest surprise came from a satellite. Neptune's large satellite Triton was a prime target for Voyager 2 not only because of its impressive size but also because it circled Neptune in a rare retrograde orbit, opposite to the spin of the planet and therefore different from all the other large satellites in our solar system. It was thought that it must be a captured body rather than one that had accreted in orbit around Neptune. Whatever its origin, Triton was revealed to have an unusual surface with variegated areas of white, gray, and salmon pink over a generally light brownish surface. The brownish tint suggested that the thin nitrogen and methane atmosphere had formed polymerized organic compounds. Active geysers or plumes of dark material, also probably organics, were seen erupting from its surface of frozen nitrogen and methane, the plumes blowing downwind for 160 kilometers (100 miles). "Fascinating," even if you are not a Mr. Spock from *Star Trek*.

Remember when SSB acting chairman Herb Friedman had written to our congressional committees in 1970 implying that a Grand Tour of the outer planets would just return pretty pictures for the news media with little scientific value? Instead, the images that had been returned by the Voyagers as they toured through the outer solar system had been astonishing—marked by one astounding discovery after another. Astronomer Ronald Schorn put it nicely when he wrote, "The Voyager missions . . . discovered, largely by means of superb images, an incredible amount of previously unimagined information about the outer planets and many of their satellites, and thereby finally demolished the mistaken idea that pictures somehow are not 'scientific.'"[8]

It was only after completing the Neptune encounter that I was able to look Jim Fletcher in the eye and reassure him that little harm had resulted after all from his terrible decision in 1971 to cancel the original J-S-P/J-U-N Grand Tours. I decided not to hold against him the loss of a Pluto flyby. It was perhaps appropriate to leave one remaining planet for today's generation to explore.

Voyagers 1 and 2 are now on their way out of the solar system and heading for the boundary between the solar wind and the interstellar wind that sweeps through our galaxy. By early 1998 Voyager 1 had already outdistanced the now-silent Pioneers 10 and 11, and with an adequate reserve of attitude-control hydrazine, it is expected to continue returning data until about the year 2020, when its RTGs will finally run out of power.

The firsts and superlatives about the Voyager missions have already filled books. The missions' science and engineering teams were exceptional, and never have I seen scientists so openly grateful to engineers for their ingenuity and dedication. (Ray Heacock has given to several audiences a one-hour talk on Voyager problems and solutions that would make any engineer swell with pride. It almost brings tears to my eyes.) Carl Sagan, too, gave tribute to the engineers: "The spacecraft phenomenally outperformed their design specifications. The bulk of our knowledge of the outer solar system has come because JPL did such a brilliant job with these extraordinary spacecraft—coming in on time, under costs, and vastly exceeding the fondest hopes of their designers."[9]

To those who would balance the budget by cutting out space exploration, Voyager from inception to Neptune flyby cost each U.S. resident only twenty cents per year, not even enough to buy a small candy bar. Not much to pay for the greatest single voyage of exploration of all time. Again, Carl Sagan said it well: "The two Voyager robots have explored four planets and nearly sixty moons. They are triumphs of human engineering and one of the glories of the American space program. They will be in the history books when much else about our time is forgotten."[10]

Expressing an opinion common among the Voyager team members, Norm Haynes said, "I consider the Voyager mission to be almost more of a success as an exploration and an adventure than as a scientific and technological accomplishment. Of course they were great technological achievements and brought great scientific returns, but I think just the adventure of going places where nobody had been before and seeing things nobody had seen before was an equal part of this enterprise. That's what excited us all, and probably excited to a certain extent the public: we were seeing things that no person had ever seen before. It was great to be present at the unveiling."[11]

Finally, space veteran Bob Parks spoke for many of us when he said, "The opportunity to have been involved in and associated with an activity such as Voyager has really been a tremendous reward to me. I was extremely lucky to be in the right place at the right time, to have been able to play a part and observe the results and see all the fantastic efforts that went into creating something like the Voyagers, and getting the results we got from them. I think that everyone who has ever touched it, was ever involved in any way, felt the same way about it. It was the kind of reward that not everybody gets in life."[12] Amen from me, every word of it. I agree wholeheartedly with my longtime friend, associate, and collaborator Bob Parks. Playing a role in making Voyager happen is the greatest thrill of my career in space exploration. How lucky can a person be in this life? I have received a number of awards over the years, but the one I value most is the Medal for Outstanding Leadership that I received from NASA with the inscription "In recognition of outstanding leadership of the Voyager Program from its very inception." Exaggerated, of course, but nice to have.

CHAPTER 8

1978

PIONEER VENUS I AND 2
ORBIT AND PROBES

By now you will have observed that no two planetary missions were sold in exactly the same way. There was no one magic formula of success. The sales teams would start out facing the same four hurdles outlined in chapter 5, but the tactics for getting over those hurdles had to be adapted to the unique selling points of individual missions and the prevailing climate of budget, scientific priorities, and competition. Getting over the hurdles would come down to "any which way you can."

With the earlier missions there was always the excitement of exploration—the allure of seeing a planet close up for the first time, of penetrating its atmosphere and even landing on its surface—but for Venus all of those firsts had been accomplished by the mid-1970s, and although the results were exciting for planetary scientists, they were not all that spectacular for the general public. (It would be more than another decade before radar mappers on the Russian Venera and American Magellan spacecraft would reveal fascinating surface structure in exciting detail.) It was not going to be easy to sell another mission to Venus.

Viewed in visible light, Venus was a pale yellow ball totally without features. Even in ultraviolet wavelengths in which cloud features were distinguishable, only large-scale swirls were apparent, without the turbulent details that were so captivating in the images of Jupiter. From Mariners 5 and 10 and the Russian Veneras we already knew that the surface was a blast furnace at 730 degrees Kelvin (850 degrees Fahrenheit) and that the dense carbon dioxide atmosphere pressed down on the surface at a pressure of more than ninety Earth atmospheres. Infrared spectroscopy had told us that the Venusian clouds were laced with droplets of sulfuric acid. There was certainly going to be no life on Venus. As for exploration, the USSR had already penetrated the atmosphere and by 1972 had successfully landed on the surface. In 1975 Veneras 9 and 10 returned pictures of the surface terrain, which was seen to be barren indeed. So we were not going to sell future missions to Venus on public appeal alone. I believed we should do more to educate the American public and their congressmen on the value of comparative planetology—specifically, what Venus could tell us about the future of Earth's atmosphere—but that was going to take time. Any immediate new Venus mission was going to have to make it into the NASA budget primarily on scientific merit. That it did.

Driven by Science

Of all the planetary missions of the 1970s golden era, Pioneer Venus '78 was the most purely scientific. It was designed by scientists for scientists and mainly sold by scientists. In my mind two atmospheric specialists, Nelson Spencer and Richard Goody, were the fathers of the Pioneer Venus project. Nelson, an energetic and blunt nuts-and-bolts scientist-engineer (degrees in electrical engineering) who specialized in the design and development of mass spectrometers for atmospheric analysis, had worked at NASA's Goddard Space Flight Center since its early years and had continually pushed to get a planetary science project at GSFC. While I was at Ford Aeronutronic in the early 1960s I had worked with Nelson and Goddard's irrepressible Bill Stroud on a joint proposal to develop a probe to enter and analyze the atmosphere of Venus. We wound up in a spirited competition with a team out of NASA Ames for the probe job. Before a winner could be picked NASA Headquarters found itself in cost-overrun problems with Surveyor and

Centaur and dropped the probe from its Venus mission plans in favor of the simpler and cheaper Mariner 5 flyby.

Whereas Nelson Spencer was strong on instrument development, the cultured, English-born Richard Goody, a professor at Harvard, was a recognized leader in the theoretical side of atmospheric physics and chemistry and a member of influential panels and committees, such as the Space Science Board. He was openly critical of the higher-level science management and prioritizing at NASA Headquarters and was more than happy to join with Nelson in planning a *real* science mission to Venus.

Although Goody and Spencer were perhaps opposite in mannerisms and temperament, their skills were complementary and they worked very well together. The Goddard Space Flight Center had long wanted to do a planetary mission (in addition to the Venus probe proposal, GSFC had submitted an early Jupiter Galactic Probe proposal that led to the Ames-led Pioneer missions to Jupiter and Saturn), so in 1967 GSFC management assembled a mission planning team under the leadership of George Levin and Chris Stephanides to support Spencer and Goody, who in turn enlisted the participation of scientists Don Hunten (Kitt Peak National Observatory and University of Arizona) and Vern Suomi (University of Wisconsin). Hunten gives the impression of being somewhat dour and speaks very deliberately, but he is highly respected among atmospheric scientists. Suomi is the opposite in personality, radiating cheerful enthusiasm, but he is equally respected, perhaps as the dean of meteorologists, and recognized as the father of the geosynchronous Earth-observing satellites that give us daily tracking of weather fronts and storms.

This strong scientific team issued a key report from Goddard in January 1969, making a strong case for a probe mission to Venus. Their work inspired twenty-one scientists of the SSB and the LPMB to perform a 1970 study of potential Venus missions. Their final report, titled "Venus—Strategy for Exploration," made a strong case for both orbiter and atmospheric probe missions to Venus. This report, known from its cover as the Purple Book, gave Code SL at NASA Headquarters enough ammunition to start waging the battle to get a strong Venus mission into the NASA budget.

Switching Centers

By this time Don Hearth had transferred from his position as director of SL to deputy center director at Goddard. I was surprised to find that my old friend Don was not all that enthusiastic about the prospects for a Venus project at Goddard. Don was a realist and knew that it was not going to be easy to sell any Venus mission. He wanted Goddard to lead NASA in Earth science and space applications and was pushing for a small applications technology satellite as Goddard's next new project.

Knowing Don's thoughts, I took the precaution of going to NASA Ames and asking the Ames director, Hans Mark, if they would undertake the Venus project should Goddard back out for any reason. Hans did not think we could sell the project, but he was looking for more work for Ames so he agreed to be a backup. A few months later it came as no surprise when Don Hearth proclaimed that Goddard could only handle one new project and they were going to bet on the small applications satellite as the one most likely to get funded. This was a major perturbation to our prospective project, but fortunately we in SL were prepared to make the switch to Ames.

To the advocates at Goddard, especially Nelson Spencer, George Levin, and the nominated project manager, Paul Marcotte, it was a tremendous blow. I really felt for them. They had worked so hard to get the ball rolling and build the necessary support in the space science community, and now the project was being snatched away from them. It was ironic that their own deputy center director was the culprit once again—earlier, while director of SL, Don had turned down Goddard's proposal for the Jupiter Galactic Probe and handed it sole source to TRW and Ames as a continuation of the Pioneer series. In spite of their great disappointment, Spencer and Levin and their Goddard team members were absolutely magnificent in their cooperation in transferring the project smoothly from Goddard to Ames. They traveled to Ames and gave the Pioneer team all of their study results so that mission planning could continue without any serious disruption. Ames named Skip Nunamaker, later to be the Pioneer Venus deputy project manager, to head their study team.

Here I should say a few words about Hans Mark, the brilliant physicist and director at Ames from 1969 to 1977 and a very controversial figure wherever he served, be it at a NASA center, as secretary of the air force from 1979 to 1981, or as deputy administrator at NASA Headquarters from 1981 to 1984. In his youth, Hans and his family in Germany had barely escaped Hitler's extermination of Europe's Jewish population, and that clearly had had a major impact on him. Survival was his driving objective in all matters. As director at Ames he worried that the center would go out of business, and I think he viewed space exploration as a national "luxury" that would be canceled in tougher economic times. On the other hand, national defense would always be necessary, so Hans wanted Ames to focus on Department of Defense work. Along this line, he kept telling Charlie Hall and the Pioneer team at Ames that Kraemer et al. would most likely fail to sell the Pioneer Venus project and that he would then abolish the Pioneer project office at Ames. Great for building morale.

Still, I very much liked dealing with Hans. Why? Because he was straightforward and communicative. He became good friends with Dan Herman, and every week either Dan or I would receive a handwritten page from him telling just what was on his mind (usually pessimistic forecasts). We called these pages "Hans-o-grams." Knowing what he was telling the troops at Ames, I would counter by sending Charlie Hall regular doses of encouragement, which helped keep the Pioneer team spirits upbeat. I contrast this with JPL, where I never knew what was on Bill Pickering's mind and where John Naugle and I seemed to be the last to know. Recently I asked Charlie Hall whether Hans Mark's pessimism had caused him any real trouble and Charlie said, "Not really, and Hans always gave me the in-house support I needed on my Pioneer projects."[1]

One more note on the fascinating career of Hans Mark. In 1981, when he was appointed deputy administrator of NASA, he made the rounds of all the NASA centers to give them some fatherly advice on how to stay busy in the future. True to form he emphasized doing more to contribute to national defense. At Goddard he was upbeat, saying the center was so diversified that it would survive even if NASA should fold. But at JPL he told them they were so tied to solar system exploration that he did not see how they could survive unless they diversified. (The JPLers were more outraged than dejected.) At other centers, such as Lewis, he just said bluntly that they were not essential to national defense and were therefore certain to go out of business. And this was the number two leader of NASA. As I said, Hans was controversial wherever he went. He really belonged in DoD, not NASA.

Organizing the Project

In January 1972 we at NASA Headquarters established the Pioneer Venus Science Steering Group to work with the Ames team to develop in detail the scientific rationale and objectives for the Pioneer Venus missions. This SSG defined twenty-four important questions about Venus (see table 2-2 in Fimmel, Colin, and Burgess, *Pioneer Venus*) that had not been resolved by any earlier Venus missions. Based on the work of the SSG, NASA Headquarters issued an announcement of opportunity (AO) for experiments on a Venus multiprobe mission. This was followed in August 1973 by an AO for a Venus orbiter mission. After intensive proposal evaluation by the SSSC, much assisted by SL's Ichtiaque Rasool and Bob Fellows, the final instrument payloads were selected on 4 June 1974.

After several proposals for earlier launch dates had failed it was now proposed that both orbiter and multiprobe missions be launched within the window in 1978. A spinning spacecraft bus was to carry four probes, one large probe surrounded by three smaller probes. The probes were to be targeted to different regions on Venus, north and south as well as equatorial, and on both day and night sides of the terminator to achieve optimum scientific sampling. All measurements from the bus, probes, and orbiter were to be coordinated to best answer the long list of scientific questions. The resulting list of selected instruments and experiments is indeed impressive:

Composition and Structure of the Atmosphere
Large probe mass spectrometer, J. Hoffman
Large probe gas chromatograph, V. Oyama

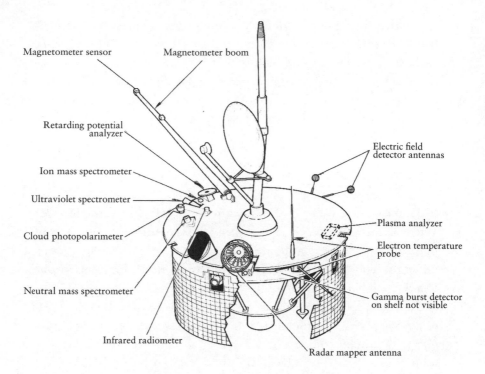

Magnetometer sensor

Magnetometer boom

Retarding potential analyzer

Electric field detector antennas

Ion mass spectrometer

Ultraviolet spectrometer

Cloud photopolarimeter

Plasma analyzer

Electron temperature probe

Neutral mass spectrometer

Gamma burst detector on shelf not visible

Infrared radiometer

Radar mapper antenna

The Pioneer Venus 1 Orbiter, a small spin-stabilized spacecraft launched on 20 May 1978 that characterized the gravity field of Venus and produced radar altimetry maps of a surprisingly diverse surface. (Courtesy NASA)

Bus neutral mass spectrometer, U. Von Zahn
Orbiter neutral mass spectrometer, H. Niemann
Orbiter ultraviolet mass spectrometer, I. Stewart
Large/small probe atmosphere structure, A. Seiff
Atmospheric propagation experiments, T. Croft
Orbiter atmospheric drag experiment, G. Keating

Clouds
Large/small probe nephelometer, B. Ragent
Large probe cloud particle size spectrometer, R. Knollenberg
Orbiter cloud photopolarimeter, J. Hansen

Thermal Balance
Large probe solar flux radiometer, M. Tomasko
Large probe infrared radiometer, R. Boese
Small probe net flux radiometer, V. Suomi
Orbiter infrared radiometer, F. Taylor

Dynamics
Differential long baseline interferometry, C. Counselman
Doppler tracking of probes, A. Kliore
Atmospheric turbulence experiments, R. Woo

Solar Wind and Ionosphere
Bus ion mass spectrometer, H. Taylor
Orbiter ion mass spectrometer, H. Taylor
Orbiter electron temperature probe, L. Brace
Orbiter retarding potential analyzer, W. Knudsen
Orbiter magnetometer, C. Russell
Orbiter plasma analyzer, J. Wolfe
Orbiter electric field detector, F. Scarf
Orbiter dual-frequency occultation experiments, A. Kliore

Surface and Interior
Orbiter radar mapper, G. Pettengill
Orbiter internal density distribution experiments, R. Phillips
Orbiter celestial mechanics experiments, I. Shapiro

High-Energy Astronomy
Orbiter gamma burst detector, W. Evans

Interdisciplinary Scientists
S. Bauer
D. Hunten
H. Masursky
T. Donahue
J. Pollack
G. McGill

R. Goody

N. Spencer

A. Nagy

G. Schubert

All of this science was to be carried out under a tight budget of less than $200 million. Quite a challenge. The instruments were selected so that each would contribute in coordination with others to answer the list of important questions about Venus. This data integration made the role of the interdisciplinary scientists especially important on the Pioneer Venus project. It was important to us in SL to have on that list the "fathers" of Pioneer Venus, Nelson Spencer and Richard Goody. Also aboard were early science planners Don Hunten and Vern Suomi (as a PI).

Tough Contract

Two contractors had expressed interest in the Pioneer Venus project: TRW teamed with Martin Marietta and Hughes Aircraft Company teamed with General Electric. Study contracts of $500,000 each were awarded to each team on 2 October 1972. With the coordinated orbiter and multiprobe missions fairly well defined, a request for proposal (RFP) was issued in June 1973. The competition was close. Some observers in the aerospace industry were betting on TRW, assuming that the proposal evaluation team at Ames would favor their longtime Pioneer associates from Redondo Beach. However, others knew that Hans Mark was an old friend and associate (on classified projects for DoD and the CIA) of Bud Whelon, who headed up the Hughes space and communications business, and they thought that relationship would tilt the scales. I saw no signs of either bias. Hughes made several right moves in their proposed designs, using a common spacecraft structure for both the orbiter and the multiprobe bus and employing a proven de-spun antenna design rather than TRW's more complex moveable platform for instruments. Furthermore, TRW was faulted for a management plan that would have charged a TRW management fee on all effort contributed by their team member, Martin Marietta. Hughes was

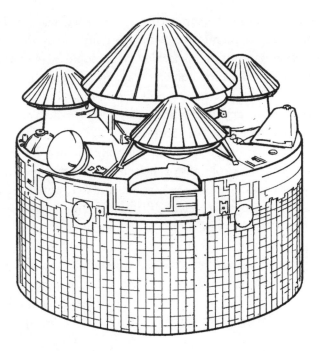

The Pioneer Venus 2 multiprobe spacecraft, consisting of a small spin-stabilized bus carrying one large atmospheric probe and three smaller probes. Launched on 8 August 1978, the probes reached Venus on 9 December 1978, examining its thick and intensely hot atmosphere all the way to the surface. (Courtesy NASA)

selected in February 1974 for negotiation of a cost-plus-award-fee (CPAF) contract.

The contract negotiations were fascinating. We at NASA were determined to keep this a low-cost project, as originally conceived. The surprise was that Hughes was willing to join in that austerity. Bud Whelon told me that he did not view the Pioneer Venus project as a source of profit—Hughes would be content to break even. Their extremely successful communications satellites were bringing in all the profit they needed to fulfill their obligations to the parent Hughes corporate office. As might be expected from the eccentric Mr. Howard Hughes, his Hughes Aircraft Company was unique in all the aerospace industry. Any profit over their specified target could be used at the discretion of management for bonuses, pursuit of new business, technology development, entertaining, any form of marketing, and so on. Their competitors complained bitterly that this gave them an unfair advantage in the competition for contracts.

Whelon said that his problem was that his engineers were getting bored with designing variations of the same old communication satellites. He was beginning to lose key people and he wanted the Pioneer Venus work because it was technically challenging and exciting and would rejuvenate his work force. In the negotiated CPAF contract Hughes agreed not only that any overruns would come out of their fee but also that it could carry into the red where they were actually paying for the overrun out of corporate funds. Really quite remarkable. And as it turned out, they indeed ran a bit into the red, but not a word of complaint. Whelon was apparently satisfied with the positive effect on his people.

A New Start

At this point the project was well organized and ready to charge into detailed design and development. There was one small hitch, however: we were not yet funded. After failing to get into NASA's fiscal year 1974 budget request we were now struggling to make it in fiscal year 1975. NASA was starting to grapple with shuttle development and the budget was getting ever tighter. Planetary Programs were getting a lion's share of the Space Science budget as Viking spending was at its peak and the effort on Voyager was building. In addition Viking was threatening to bring down the whole house of cards. As discussed earlier, we were facing formidable technical challenges in almost every aspect of the overly ambitious Viking Mars Lander. This was a billion-dollar project (about four billion in 1997 dollars), and a failure would shake the entire agency and certainly doom the prospects for any new planetary missions for many years to come. I found I was running all over the country with the Viking managers giving pep talks to all the struggling subcontractors (during 1974 I was on travel status 104 days).

I was certainly neglecting Pioneer Venus. So I met with my deputy and chief scientist, Ichtiaque Rasool, and my other right-hand man, Dan Herman, and told them that I would sweat out Viking but wanted them to concentrate their considerable talents and energy on selling Pioneer Venus. They did a tremendous job. Ichtiaque marshaled all the planetary scientists, and Dan helped focus the sales force at Hughes and General Electric. Senior atmospheric scientist Tom Donahue from the University of Pittsburgh even gave a stirring sales pitch directly to OMB management, stressing the potential benefits of Venus science to Earth meteorology. Pioneer Venus finally made it into the NASA budget request for fiscal year 1975, and in August 1974, thanks to superb mission planning, effective lobbying by scientists and contractors, the growing appreciation for comparative planetology, and the relatively low cost of the mission, Congress approved our new start. We were off and charging toward a pair of launches in 1978.

Developing the Spacecraft

Whelon picked one of his rising young stars, Steve Dorfman, to head up the Hughes project team. Steve was not at all the typical project manager, who is usually a forceful type, such as Ames' Charlie Hall, Langley's Jim Martin, JPL's Gene Giberson, Martin's Walt Lowrie, and Boeing's Ed Czarnecki. Steve, who always dressed like an aspiring Wall Street executive, was an exceptionally composed and well-spoken young man. Obviously highly intelligent, he happily turned out to be an effective project manager, and his team members respected his judgment. Other key people on the Hughes Pioneer Venus team included John Bozajian as spacecraft engineer, Mal Meredith as test engineer, Jack Fisher as systems engineer, and Leo Nolte in charge of the multiprobes.

There were some interesting aspects to the spacecraft design, especially the multiprobes. To withstand the high loads and elevated temperatures of entry into the dense Venusian atmosphere, the probe spherical pressure vessels needed to be made of titanium, then a rather exotic metal. We were a bit shaken to discover that only one tiny company that no one had heard of, Arcturns Manufacturing Company in Oxnard, California, was willing to attempt to forge the shells from titanium billets. To everyone's relief, they delivered beautiful shells.

Instruments had to view through these pressure vessels through special windows that could withstand high pressure, high temperatures, and the corrosive action of sulfuric acid. The ideal window material for instruments viewing in the optical and ultraviolet bands was sapphire. No problem—industrial-grade sapphire could be manufactured and was readily available. However, sapphire would not pass the infrared wavelengths required for the large probe infrared radiometer. The only suitable material for that window was diamond, and it had to be a huge three-quarters of an inch in diameter. Oh, boy, I thought, wait until the news media picks this up. We will be a sure candidate for Senator Proxmire's infamous Golden Fleece Awards. Fortunately, the specific diamond that had the transmissibility properties we needed was a type 2A, which is almost perfectly pure but has a slight brownish cast that renders it less than ideal for jewelry and therefore classifies it as relatively low-cost industrial diamond. Nevertheless, a diamond of the size we required is quite rare. We had to commission a dealer to visit his contacts in South Africa to search out a suitable stone. He came back with two good candidates, and the total cost was about sixty thousand dollars, not enough to attract the attention of Senator Proxmire. In fact, we were able to avoid twelve thousand dollars in U.S. import taxes by arguing that the diamonds were only "in transit" from the United States to Venus.

During the design phase it was becoming increasingly difficult to stay within the launch weight limitations of the planned Delta launch vehicles. This was forcing overly sophisticated designs and the use of exotic materials that were driving the cost up. With a helpful push from George Low we switched over to our standard planetary workhorse, the Atlas-Centaur. It cost more than the Delta but would save overall costs.

There were plenty of development problems to keep all of the Pioneer Venus engineers and PIs busy. Probably the one that most concerned Charlie Hall and Steve Dorfman was the repeated failure of the large probe parachute. This parachute was absolutely essential if the probe were to make measurements during its descent through the Venusian clouds, yet it kept disintegrating into shreds during drop tests. To better study the parachute deployment it was mounted in Ames' low-speed forty-by-eighty-foot wind tunnel. Here the parachute gores failed to open. So the air velocity was taken down to a mild twenty-mile-an-hour breeze and the sturdy Ralph Holtzclaw went in and stood by the parachute assembly to observe its dynamics. It still failed to deploy—Ralph pulled on it but could still not get the gores to deploy. He found that even at just twenty miles an hour the wind loads were enough to hold the gores together. The system was redesigned and rushed to White Sands in New Mexico for another drop test from a balloon at ten-mile altitude. Total failure. And time was running out. Studies of photographs and the wreckage indicated a fault with the test apparatus rather than the parachute itself. Another test vehicle was built in haste and the drop test repeated. Success, at last! Fortunately, Charlie Hall's hair was already white, so we saw no permanent impact on him from this adventure.

A Congressional Crisis

The problems were not all technical. Through no fault of our own we were suddenly in deep trouble with Congress. Bill Lilly, the NASA comptroller, was always searching for ways to hide a few reserve dollars in the NASA budget to cover inevitable funding problems. (He was able to rescue us from Viking overruns more than once.) He knew that we had the Pioneer Venus Project under tight budget control and were determined not to overrun our budget, so for NASA's fiscal year 1976 budget request he padded our Pioneer Venus budget by several tens of millions of dollars as his own hidden reserve.

The budget got past our key Authorization Subcommittee in the House and went on to the Appropriations Subcommittee chaired by Rep. Joseph Boland. In those days Appropriations was not usually a hurdle—any cuts usually came only from Authorization—but Joe

Boland had made a commitment to add $10 million to the appropriations for Housing and Urban Development (HUD). He asked his able and experienced senior staff member, Dick Mallow, to find $10 million from somewhere in the NASA budget request. Dick readily noticed that Pioneer Venus seemed to be spending behind its requested budget, so he reasonably concluded that he could take away a paltry $10 million and do no one any harm. In fact, there was so much apparent slack that Boland went overboard and wound up cutting Pioneer Venus from $57 million to $9.2 million, a reduction well beyond Lilly's hidden reserve. So here we were on the Pioneer Venus project, cruising along right on our budget plan and meeting all major milestones, yet suddenly we were faced with slipping the launches from 1978 to 1980, which would have run our costs way over our $200 million ceiling.

Bill Lilly did not want to admit to Boland that he had any reserve in the NASA budget, so we were forbidden to tell Mallow the true budget story. In my office Rasool and Herman responded immediately to emphasize the importance of the Pioneer Venus mission. They helped organize the testimony of the nation's most eminent meteorologists and climatologists to educate the congressional committees on the importance of studying the atmospheres of Venus and Mars to better understand and predict the future climate changes on Earth. Organizations such as the Space Science Board lent their support. I tried to take their message to the American voters by writing the widely published article on comparative planetology referred to in chapter 6. The following are a few excerpts relevant to Venus:

> Neighboring worlds of the solar system are undisturbed models of our planet, subject to the same universal laws of physics and evolution. Moreover, they are "galactic time machines"—mirrors of our past and future. Locked within these spheres are answers to some of the most critical questions facing us.
>
> Take Venus and Mars, for instance. These two Earth-like planets share a common origin with our world, but each appears to be in a different state of evolution. On Venus, which is closer to the Sun and warmer than Earth, water has been driven off and carbonates in the crust broken down to create a thick, poisonous atmosphere of carbon dioxide that traps and holds solar heat. In effect, Venus has become a huge uninhabitable greenhouse. Is this the ultimate fate of Earth?

Spacecraft surveillance of Venus has contributed to Earth's weather picture as well, confirming a major theory dating to 1735. The British astronomer George Hadley, trying to explain and chart the steady trade winds essential to sailing ships of that era, contended that a solar-driven circulation of air from the equator to the poles powered Earth's weather.

The theory has made sense to meteorologists ever since, but proving it was another matter. Even from the vantage point offered by today's weather satellites, the cloud patterns in Earth's atmosphere are so confused by storms and cyclones that the overall Hadley-cell circulation is obscured completely.

The answer was provided two years ago when Mariner 10 darted past Venus en route to Mercury. Closeup photographs sent back by the spaceship showed great swirling cloud patterns sweeping poleward from the Venusian equator exactly as theorized for Earth centuries ago. Enthusiastic meteorologists predict that the Mariner 10 photographs will stand alongside Hadley's classic drawings in every meteorology textbook published for the next one hundred years.

The planets also have alerted us to a new, insipid danger in our atmosphere—ozone depletion. Earth's ozone shield is vital to all life since it absorbs most of the Sun's deadly ultraviolet radiation, leaving us with only a sunburn. Without ozone, life would be impossible. Even a slight decrease in ozone protection would cause a significant increase in skin cancer among humans.

Mariner 10's 1974 dash past Venus augmented telescope results to reveal a shocking fact—the Venusian atmosphere contains measurable amounts of chlorine compounds but no ozone. Chlorine, through a chain of catalytic reactions triggered by sunlight, can become a voracious consumer of ozone.

Compounds containing chlorine are being exuded in large quantities into our skies by aerosol spray cans and a number of industrial and agricultural processes. Intensive investigations of atmospheric chlorine pollution are now under way. Here, again, spacecraft surveillance of a world millions of miles away has alerted us to a condition that might have critical consequences here.

I hope these articles did some good, but they were not published in Sunday supplement magazines until May and June 1976, too late to help the immediate congressional crisis in 1975.

Having made his cut, Boland did not want to lose stature by backing off. Pioneer Venus had to work its way back, committee by committee. In July 1975 funds for Pioneer Venus were restored by a Senate subcommittee; in August the full Senate approved the requested funding. In September 1975 the Senate-House Conference Committee restored all but $1 million of the Pioneer Venus funds and we were past the crisis.

This seemingly minor move by Bill Lilly to get some needed budget reserve was to have long-lasting consequences. Boland was angered by perceived NASA double-talk and determined to have a stronger say on the NASA budget in the future. In 1977 he tried to kill our proposed Jupiter Orbiter-Probe Project (JOP, later renamed Galileo), and on 19 July he was overruled in his attempt in a vote on the floor of the House, an unprecedented defeat for a committee chairman. Now Boland was really mad. He gave NASA fits for years after that and in the process managed to wrest power from the House Authorization Subcommittee. Today it is the appropriation committees that control the NASA budget, with the authorization committees trying to steer without any firm connection to the front wheels. This was a major shift in power, and it goes back to an innocent little budget padding on Pioneer Venus in 1975.

On to Venus

In February 1978 the spacecraft were shipped to the cape, where Ed Muckley of NASA Lewis and John Neilon and his crew at NASA Kennedy had a pair of Atlas-Centaurs ready for payload integration. On 20 May 1978 the orbiter was launched right on schedule on its way to Venus. With only a two-day hold for ground equipment, the multiprobe was sent speeding to Venus on 8 August 1978.

Mission operations at Venus went just beautifully. Orbit insertion had its moment of suspense heightened by the requirement to fire the retro motor while the spacecraft was behind the planet, but it emerged in a fine orbit. The multiprobe required the simultaneous tracking and signal detection from five objects—the bus, the large probe, and the three small probes. The DSN stations at Goldstone and Canberra engaged in a friendly competition to see who could lock on to the signals first. They tied. To quote Charlie Hall, "To have them come on

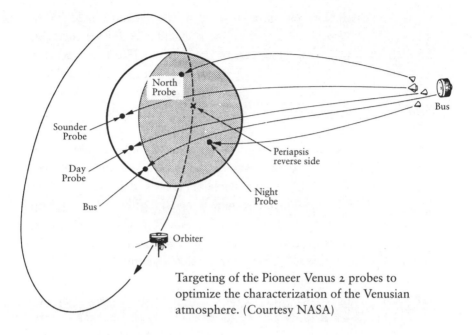

Targeting of the Pioneer Venus 2 probes to optimize the characterization of the Venusian atmosphere. (Courtesy NASA)

within a split second of the times they were supposed to, and particularly to have the ground stations lockup, was quite an achievement. I think that the lockup of the four probes was probably one of the most difficult tasks that the Deep Space Network has ever had to deal with."[2]

As it turned out, the month of December 1978 saw ten separate scientifically instrumented vehicles conduct an intensive exploration of the Earth-twin planet Venus. From Pioneer Venus there were the orbiter, the bus, and four probes, and the Soviet's Venera 11 and 12 contributed two flybys and two landers. A formal exchange of all scientific data was arranged through the joint U.S.-USSR Working Group on Near-Earth Space, the Moon, and Planets. The result was an outpouring of new knowledge and understanding of our closest planet neighbor in space. I will not even attempt to summarize it here; the reader is instead referred to the excellent publication *Pioneer Venus* by Richard Fimmel, Lawrence Colin, and Eric Burgess. I guess what surprised me most was the radar altimeter returns from the orbiter. Somehow, perhaps from the Venera Lander pictures, I expected Venus to be pretty flat. Yet here were great raised plateaus, one as large as the continental United States, and a rift valley every bit as deep as the spectacular Valles Marineris on Mars, both of which make our Grand Canyon look like a mere scratch on the surface. These findings made it much easier to sell the later Magellan radar imaging mission to map the topography of Venus in fine detail.

After leading the Hughes team to this grand mission success I thought that Steve Dorfman might well have been bitten by the exploration bug like so many of the rest of us. I sounded him out on the possibility of serving at NASA Headquarters. He looked at me like I must be mad and politely declined. He knew full well that he was headed up the lucrative management ladder at Hughes and he was quite correct. Before long he was chairman of the Hughes Telecommunications and Space Company, leading the world in satellite communications.

There is no way I could list all the heroes of the Pioneer Venus Project. Charlie Hall's small team at Ames tends to be overlooked, but they should not be. Skip Nunamaker added his energy as Charlie's deputy, Larry Colin did a terrific job as project scientist, Ralph Holtzclaw excelled as Spacecraft Systems manager (and the guy who stood in the wind tunnel to struggle with the faulty parachute), Joel Sperans accomplished wonders in getting the multitude of science instruments delivered on time and within budget, and Ralph Hofstetter smoothly managed the complex Mission Operations. Within Code SL at NASA Headquarters the day-to-day essential support to Pioneer Venus was ably supplied by the widely experienced Fred Kochendorfer as program manager assisted by his deputy, the dependable and gentlemanly Paul Tarver, with quiet but highly respected atmospheric specialist Bob Fellows as program scientist. These and many others can be proud of the contribution their creation, Pioneer Venus, made to the understanding of planetary processes, including the complex meteorology and climatology of our home spacecraft, Earth.

EPILOGUE

FROM GLOOM OF NIGHT
TO NEW LIGHT OF DAWN

What a fabulous string of missions! Looking back, it is no wonder that the 1970s has been called the golden era of planetary exploration. Over a span of just eight years Americans launched pioneering robot space machines that returned first-ever high-resolution, full-color closeup views of every planet in our solar system except remote Pluto. The scientific harvest was rich indeed. The planets were full of surprises, seeming to delight in failing to match the predictions of planetary experts, and the planets' satellites were even more astounding. Slowly rotating Mercury was found to have an improbable magnetic field and even a tenuous atmosphere of helium. Venus' sulfuric-acid clouds contained upper layers that rotated the planet at enormous velocity, and the Venusian surface featured dramatic high plateaus and deep valleys. One volcano on Mars soared over 21.7 kilometers (70,000 feet), and a Martian Grand Canyon 6.4 kilometers (4 miles) deep stretched for almost 4,800 kilometers (3,000 miles) and showed irrefutable evidence of past massive water flows and conditions conducive to the evolution of living organisms. Enormous Jupiter more than lived up to expectations, even revealing a ring, but it was its mini–solar system of large Galilean satellites that astounded the world:

Io had a whole family of active volcanoes spewing plumes of brilliant orangish red sulfur, and Europa was covered with smooth ice that showed every indication of covering an ocean of liquid water. Saturn's rings were fantastically beautiful in closeup, composed of hundreds of delicate sparkling icy ringlets. Its largest satellite, Titan, boasted a very thick atmosphere that promised a deposition of complex organics on its surface. Uranus had a weirdly tilted magnetic field and perhaps the strangest satellite in all the solar system in Miranda, with its unbelievably tortured terrain. Not to be outdone, Neptune showcased its own large satellite, Triton, which was not only orbiting in a rare retrograde direction but had a variegated pastel surface featuring organic-spewing geysers. We now had a large family of widely differing planetary bodies to compare with Earth and to help with deciphering Earth's evolution, both past and future.

You can understand that everyone associated with these missions was on an intellectual and emotional high. We had reeled off twelve missions in a row, launching at least one every year for eight years, and every one was a spectacular success, revealing unexpected wonders in our solar system. Carl Sagan spoke for all of us when he said many times how fantastically lucky we were to be participants in this unique period in history. Even as early as 1974 he enthused, "Clearly the best time to be alive is when you start out wondering and end up knowing. There is only one generation in the whole history of mankind in that position. Us."[1]

Surely after such a record of mission success with such spectacular results and mostly within tight project budgets, the administration and Congress would have swiftly approved the planned continuing program of planetary exploration. Historians may puzzle over the fact that just the opposite occurred. The pace of exploration not only slowed, but planetary launches came to a complete halt. Unbelievably, America did not launch another planetary mission for the next eleven years.[2] Truly amazing. Almost beyond imagination, in fact. What happened? Bruce Murray and John Naugle, among others, put the blame entirely on NASA's space shuttle. The scramble to cover the growing costs of shuttle development was indeed the largest factor, but there were other contributors as well.

Self-Inflicted Obstacles

Planetary scientists did not help their own cause. Flushed by their success, especially with the ambitious Viking Mars Landing missions, they felt that follow-on funding would come in generous amounts. Viking, however, had been a real gut buster that took enormous effort to pull off. We had barely snatched success from the jaws of ignominious disaster. I proposed more modest follow-on missions. One that I especially liked was a Mars roving mission. We had a spare Viking lander in storage at the Martin Marietta plant in Denver, mostly assembled and already paid for. The engineers at MSFC had developed some clever little track assemblies for potential future lunar roving vehicles. If we mounted one of those tracks on each of the three legs of the Viking lander we would have a very competent Mars roving vehicle at a bargain price, estimated at $300 million for the entire mission. With the lander's RTG power the rover could travel for years, covering many hundreds of kilometers. We could analyze the soil and rocks and send back daily panoramic images to Earth to help plan the next day's traverse. I could picture millions of viewers around the world glued to their television sets each day waiting for a view into a new canyon spotted in yesterday's images. Here was good science that also would have great public appeal. The head of the Viking Lander Imaging Team, Prof. Thomas "Tim" Mutch, was so intrigued with the concept that he agreed to leave his tenured university post and come to Headquarters to replace a departing Noel Hinners as NASA's associate administrator for Space Science and Applications.

In spite of Tim's efforts, and the respect his fellow scientists had for him, we were unable to gain any kind of consensus among planetary scientists for the rover mission. The majority felt that nothing would be denied them after the Viking success and that what they ultimately needed was to return Mars surface samples to Earth. That was indeed true, but a sample return mission using 1970s technology would have cost billions of dollars and there was absolutely no way I or my very capable successors could get that into the NASA budget, even with Tim Mutch's full support. So scientists' delusions of grandeur killed any immediate follow-on Mars mission. Tim was very frustrated, and it is especially sad during this low point in his career in 1981 that he tragically died on a mountain climbing expedition in the Himalayas. The best we could do was name the operating Viking 1 Lander on Mars the Tim Mutch Station in his memory.

The Mars rover was not the only proposed bargain mission that caused planetary scientists to turn up their noses. After the launch of Mariner 10 in 1973 we had a complete test spacecraft in good condition. A study showed that it would make an excellent spacecraft for a comet flyby mission. Even the pointing geometry of the scan platform was just right without modification, and the instrument payload was appropriate. As a result, mission price would be very modest. I thought this proposed mission would garner good scientific support. One of our major goals, as defined by the NAS, was to understand the origin and evolution of the solar system, and the one place to find the primordial material from which the solar system formed was in the comets. All other bodies had been grossly changed and differentiated by heat, compaction, impact, volcanism, and a long list of dynamic processes, but the original gases of the solar nebula had frozen into comets on the outer fringes of the nebula as the system cooled. Our one potential source for study of primordial material was in those comets that had been deflected inward within range of our spacecraft.

Did our planetary scientists leap at this bargain opportunity? I am afraid not. Most gave priority to Mars or Jupiter. Those that were especially interested in comets pushed for a much more expensive comet rendezvous mission so that a sample of the comet nucleus could be analyzed directly—a worthy objective but much too expensive to be saleable at that time. A rendezvous with even a short-period comet, much less speedy Halley's Comet, required a high level of propulsive velocity, beyond the practical limit for conventional rockets and the relatively heavy spacecraft of the 1970s. However, the required high velocities could be achieved with more-advanced solar electric propulsion (SEP) using the Sun's energy converted into electricity to accelerate ionized gases such as xenon to very high exhaust velocities. Such a system had been proposed by von Braun's senior scientist, the gentlemanly scholar Ernst Stuhlinger, back in the 1950s and demonstrated to be feasible by experimental work at Hughes, TRW, JPL, and LeRC.

NASA's Office of Aeronautics and Space Technology had done their part by flying the SERT 1 and SERT 2 ion-propulsion test spacecraft, but jockeying among JPL, LeRC, and MSFC for the NASA lead responsibility contributed to the failure to secure a go-ahead for an operational system.[3] So comet rendezvous was relegated to future decades. Can't win them all, I suppose, and we certainly had a beautiful series of missions underway during the golden era. Still, our failure to sell a comet mission was for me a deep disappointment during the otherwise marvelous 1970s.

In 1973 we were visited at NASA Headquarters by delegations from both Japan and the European Space Agency. I gave them both a pep talk on comet missions, expressing my belief in the scientific importance of the comets, but explaining that NASA was heavily committed to expensive projects to Mars (Viking) and the outer planets (Grand Tour) so it would be some years before we could start a comet project. Here was a chance for them to undertake a mission that would be a historic first and a great boost for their own space exploration programs. The Japanese reaction was interesting. They would nod and agree but then ask, "But if a comet mission is so important then why does NASA not do it?" I would repeat again that we were heavily committed to Mars and the outer planets and it would take us a while to get to comets. There would be a pause and then they would come back with the same question all over again. I was pleased that both parties eventually launched comet missions. When excitingly volatile Halley's Comet made its infrequent (every seventy-six years) return to the Sun in 1986, that great space power, the United States, sat on the sidelines while an armada of five spacecraft from ESA, Japan, and the USSR visited Halley and returned spectacular and surprising images and data, confirming astronomer Fred Whipple's 1950 prediction that comets are like "dirty snowballs." Real excitement to view for the first time the almost black comet spewing out great jets of steam, but it served us right to be little more than just a bystander.[4]

Before leaving this topic, it would be unfair not to note the vigorous fight for a Comet Halley mission made by Bruce Murray, who succeeded Bill Pickering as director of JPL in April 1976. Bruce challenged his JPL mission designers to come up with exciting, colorful missions, which he called "Purple Pigeons," rather than scientifically justified but more mundane "Gray Mouse" candidates. The Purple Pigeon he selected for an all-out push was a solar sailing mission to rendezvous with Halley's Comet. Now this was about as purple as you could get. Halley is in a highly inclined elliptical orbit with a close approach to the Sun, so that its velocity at approach is very high. In addition, it is in a retrograde orbit so that Earth's velocity around the Sun is a penalty rather than a help to the launch vehicle. As a result, even a simple flyby mission would have a comet encounter velocity an order of magnitude faster than any other mission to date. To overcome that velocity and achieve the desired rendezvous with the comet using chemical propulsion would take at least a Saturn 5 launch vehicle, and the immense Saturns were long since out of production.

To surmount these extreme velocity challenges, Bruce and his imaginative planners proposed to employ the bold but untried technique of solar sailing, utilizing the pressure from the Sun's radiations against a huge sail of thin plastic. JPL organized a mission design and sales team headed by superenergetic and hard-driving Louis Friedman. They and the OSSA people at NASA Headquarters gave it a good try, eventually backing off to simpler Halley probe and flyby proposals, but came away denied as time ran out. No Purple Pigeon was sold, and a frustrated Bruce Murray returned to his professorship at Caltech in July 1982.[5] One of the disappointed comet scientists at JPL during that time was Dr. Wesley Huntress, who, when he became NASA's associate administrator for Space Science in the 1990s, was able to direct more attention to comets.

The Shuttle Clobbers Planetary Missions

Meanwhile, the space shuttle came storming into the NASA budget picture. The ultimate rationale for the shuttle was articulated by Wernher von Braun, whose major space goal was to send humans to Mars. Before doing that one needed an orbiting space station to develop life-support systems and techniques to permit space voyages of several years, but one could not support years of continuing space station operations using enormously expensive Saturn 5 launch vehicles. Costs would have to be brought down by employing recoverable launch vehicles that could be used over and over again. In other words, a space shuttle.

Unfortunately, the road to a shuttle soon became rocky. The Nixon administration and its Office of Management and Budget set a constrained ceiling on the future NASA budget. It was von Braun himself, drafted from MSFC to Headquarters to help sell the shuttle, who pointed out that there was no way a fully recoverable vehicle could be developed under the restrained NASA budget. So the design had to be compromised with solid-propellant boosters (only partially reusable) and a large throwaway external tank, making the economics marginal. James Fletcher, newly aboard as NASA administrator, was convinced by advisors such as Willis Shapley that manned space flight was absolutely essential to continued public support for NASA, so he gave the shuttle program top priority. To Jim's credit, he insisted that NASA would only undertake the project if the NASA budget went up yearly by at least the amount of inflation. That was the agreement for the "golden handshake" Fletcher had with President Richard Nixon at San Clemente on 5 January 1972. Fletcher proved to be naïve. The very next budget from OMB reduced the allocation to NASA, and the budget struggle of the next decade began. There was no longer room in the reduced NASA budget for both development of the shuttle and a vigorous space science program. And the NASA administrator felt that he absolutely had to continue to support manned space flight, constrained budget or not.

The shuttle had an immediate and crushing impact on our plans for exploring the outer planets. I had high hopes of selling two series of outer-planet spacecraft. First, a family of the simple and long-lived Pioneer spacecraft. As spinners, they were ideally suited for delivering atmospheric probes and for making particles and fields measurements, for which the PIs always wanted complete 360-degree scans. The second series would be three-axis stabilized Mariner-class spacecraft ideally suited for high-resolution imaging and for aiming spectrometers and other pointing instruments. By having a series of standardized spacecraft, costs would be greatly reduced; progressive exploration of the outer planets could be conducted systematically and within a reasonable budget. With shuttle development costs rising, though, no one in OMB or Congress was willing to commit to multiyear funding beyond the upcoming annual budget.

In one of my regular meetings with Bob Parks, director of all spaceflight projects at JPL, I told him a commitment to a series of outer planet missions was not realistic in the current budget climate, and that we had a better chance of selling just a single mission. There was very strong support among scientists, including Charlie Townes and his colleagues on the SSB, for a Jupiter orbiter mission, and some of the most influential planetary scientists, such as Richard Goody and Tom Donahue, were atmospheric specialists who very much advocated a Jupiter atmospheric probe. We would be able to get good scientific support for a combined Jupiter Orbiter-Probe. The people at NASA Ames already had the probe pretty well defined and had solid analysis and test data to confirm that surviving the extreme high speed entry into the Jovian atmosphere was technically feasible. Of course, Ames assumed that the probe would be delivered by one of their spinning Pioneer spacecraft. If we could only sell one spacecraft, however, I wanted the pointing capability of a Mariner rather than a spinning Pioneer. I suggested to Bob Parks that JPL consider a hybrid spacecraft, basically three-axis-stabilized but with a spinning section for the probe and particles and fields measurements.

The initial reaction from JPL was understandably negative. A spinning section normally required slip rings to carry electrical data to the three-axis bus for transmittal to Earth, and slip rings were not noted for long life. Still, as I reminded Bob, Hughes Aircraft Company had experienced excellent success and many years of lifetime with the despun antennas on their spinning communications satellites. JPL eventually convinced themselves, and we were able to get JOP (later renamed Galileo) into the budget for launch in January 1982. As our last new start before I left NASA Headquarters (and, as it turned out, the last planetary new start for a decade) I felt a special attachment to Galileo and closely followed its progress, which turned from nervous into downright agonizing.

First of all, the JPL-managed Seasat satellite (Gene Giberson as project manager) had its life cut short by a failure of its slip rings, and I began to feel that it would be all my fault if slip rings doomed Galileo. Problems coming from the shuttle soon made that worry seem trivial, though. Jim Fletcher found that with the shuttle he had grabbed a bull by the tail and did not dare let go. Although the shuttle was really intended as a means of servicing a space station as a necessary step toward sending humans to Mars, Fletcher found that he was now forced by OMB into the uncomfortable position of trying to justify it on strictly economic benefits. To achieve real savings meant the launch rate and utilization of each shuttle had to be maximized. Thus, although only the Large Space Telescope among all the planned NASA payloads required the shuttle, Fletcher decreed that *all* of NASA's major spacecraft were to be launched on the shuttle.

We were told that we had to use the shuttle, rather than the more than adequate Titan-Centaur, to launch Galileo. Jim Fletcher had personally assured our planetary scientists that the shuttle would at least equal the launch capability of the Titan-Centaur. He proved to be wrong. From shuttle orbit the Galileo spacecraft was to be boosted by the solid-propellant IUS (which initially stood for interim upper stage, but later was changed to the less-interim-sounding title of inertial upper stage), but predicted IUS performance kept dropping. Then shuttle development delays slipped the planned Galileo launch date from 1982 to 1984, a less favorable year. Galileo's weight had to be reduced drastically, forcing the spacecraft to be split into two parts for two separate launches: a major redesign. Meanwhile, in Congress, the powerful Rep. Joe Boland had been increasing the pressure on NASA to utilize its Centaur stage in the shuttle, which would greatly increase its capability over the IUS. Finally, on 15 January 1981, Bob Frosch, appointed by newly elected Democratic president Jimmy Carter to replace the Republican Jim Fletcher as NASA administrator, caved in to Boland's pressure and agreed to the shuttle-Centaur combination for Galileo.

Alleluia, we were back to one spacecraft again. It would take time, however, to adapt the Centaur to the shuttle, and the launch date was slipped once again, this time to 1986. By early 1986 the Galileo spacecraft was at Cape Canaveral for launch preparations. The launch crew was hurrying preparations for a January shuttle Challenger launch—

they wanted as much time as possible to get ready for the ensuing launch of Galileo with its potentially hazardous hydrogen-oxygen Centaur stage going into the shuttle cargo bay for the first time.

On the cold morning of 28 January 1986 Challenger was launched. One look at the launch replay on the television monitor and I knew the Challenger crew was doomed, and then my next thought was, Goodbye Centaur, goodbye Galileo launch. After a horrendous accident such as Challenger there was no way NASA would ever put a potentially explosive hydrogen-oxygen stage in the shuttle. Galileo was without a ride again.

Well, in spite of all the hurdles thrown its way by the shuttle, Galileo did make it to Jupiter. A launch in 1989 was made feasible by employing gravity assists from Venus and two Earth flybys. This required a six-year journey rather than less than two years for a direct flight. Design modifications had to be made to the spacecraft to enable it to handle the increased solar heating from going inward toward the Sun to reach Venus. In hindsight the added shipping of the spacecraft back and forth between JPL and the cape, coupled with the added years of storage, probably degraded the dry lubricant and prevented Galileo's high-gain antenna from fully deploying, rendering it useless. That the Galileo mission recovered and succeeded so well, that it ever in fact was launched, is a tribute to the entire Galileo team. My own special hero is the irrepressible John Casani, the Galileo project manager who stuck with the project through thick and thin, through repeated redesigns, and through years of launch delays. I marvel that John was able to rouse his team from despair time after time and keep them motivated over an endless decade to achieve a highly successful mission. John was successively promoted to direct all of JPL's flight projects and then to become JPL's chief engineer.

So Bruce Murray and others are no doubt correct in putting much of the blame on the shuttle for causing Planetary Programs great woe and for being a major cause of the eleven-year lapse in the launch of planetary missions from 1978 to 1989. Murray in turn puts the burden on Jim Fletcher for making poor decisions on the shuttle. I am less inclined to do that. I liked Jim and deeply regretted his untimely death in 1992. As NASA administrator he did not stand on rigid formality and strict management channels of communication; at least once a week he would catch me in the corridors or lunch room at NASA Headquarters to get a personal update on how we were progressing with technical problems on our planetary spacecraft. He came to some of our planetary launches, and whenever we had a planetary encounter Jim was there at the SFOF mission control center at JPL—most of his predecessors and successors cannot claim that level of interest. As a NASA administrator, though, Jim was not in a class with Jim Webb. (But who was?) He was just not a great salesman, and in my opinion selling is the most important role of NASA Headquarters. And surprisingly for an experienced aerospace executive, Jim seemed rather naïve. He sometimes accepted without critical questioning optimistic predictions from his program people or else relied completely on his deputy, George Low, and appeared easily swayed by opinions from strong-minded staff members such as Bill Lilly and Willis Shapley. Yet I never questioned his sincerity and good intentions. It would be great if we could combine the "nice-guy" attributes of Jim Fletcher with the tough decision-making talents of his later successor, Dan Goldin.

While the shuttle was clobbering the prospects for most space science projects, one extremely valuable project was spared. The Large Space Telescope, later named Hubble after the famed astronomer, needed the shuttle, and vice versa. The LST, designed for reservicing and instrument replacement by manned visits via the shuttle, was so large that it would fill the entire shuttle payload bay. In spite of difficulties and major cost overruns, it sailed through the tough years of shuttle development. The importance of pairing the Hubble with the shuttle was vividly demonstrated when a repair in orbit was necessary to correct a faulty mirror. Subsequent servicing via the shuttle is keeping Hubble up to date, and there is always the option to use the shuttle to return the telescope to Earth for major refitting.

Resurrection of Planetary Missions

As funding for shuttle development wound down, a narrow opening for worthy new starts opened up. Code SL and JPL were able to get two significant new planetary projects started—Mars Observer and Magellan. Mars Observer, launched in 1992, experienced a propulsion system failure as it was about to go into orbit around Mars, the first failure of a planetary mission since the launch misfortune of Mariner 8 way back in 1971. Magellan, on the other hand, launched 4 May 1989, was a spectacular success, returning a detailed topographical map of essentially the entire surface of Venus, revealing at a resolution of a few hundred feet a much more varied and spectacular terrain than I would ever have imaged. I give Dan Herman a great deal of the credit for selling that new start. Dan had worked with new synthetic aperture radar (SAR) techniques while at Northrup, knew what spectacular images SAR could produce, and had been pushing for a radar mapping mission to Venus from the time he joined us in Code SL in 1970. I was not too supportive. Radar missions are difficult, requiring a tremendous amount of data to be returned to Earth, and converting the received data into images is also a daunting task. Besides, I really did not think the surface of Venus would be all that interesting. I was wrong. I learned that the planets never let us down. The mission might still be just a distant dream if Dan Herman and the succession of directors of SL had not kept pushing the planning and technology development through the difficult years until an opening appeared in the NASA budget.

Later it was possible to get a start for a joint U.S.-European Saturn Orbiter–Titan Probe mission named after the astronomer Cassini. Permission was granted to go back to the potent Titan-Centaur launch vehicle, but the spacecraft is so large that it takes two gravity assists at Venus, one at Earth and another at Jupiter, to send Cassini on its seven-year journey to a Saturn arrival on 1 July 2004. This will be the last of the large-spacecraft missions to the planets for the foreseeable future. There is now a national commitment to budget deficit reduction, and it appears NASA will be constrained by an extremely tight budget for years to come. Those of us who dream of that inevitable day when the first humans set foot on Mars are going to have to be very patient.

Happily, the Planetary Programs people were among the very first to recognize the coming wave of austerity and need for downsizing. Under the leadership of Wes Huntress at NASA Headquarters and Ed Stone at JPL, missions to explore the solar system were designed employing the latest technology to accomplish significant exploration employing small and relatively inexpensive spacecraft. When Dan Goldin came aboard as NASA administrator in April 1992 with his theme song of "faster, better, cheaper," Wes Huntress was able to respond immediately with viable-candidate mini-missions to Mars, asteroids, the Moon, comets, and even unexplored Pluto. Goldin was enthusiastic—someone was proving him right, that it was indeed possible to move smartly ahead in space science and do it faster and cheaper, and perhaps even better. Already at the time of this writing the Near-Earth Asteroid Rendezvous (NEAR) spacecraft has rendezvoused with the asteroid Eros, the Pathfinder spacecraft (now named the Carl Sagan Memorial Station) has sent back color images and data from its landing spot on Mars while its tiny Sojourner companion has merrily roved around analyzing a variety of Martian rocks, and Mars Global Surveyor is in Mars orbit and already producing improved maps of the surface of Mars. Modestly scaled missions are being planned and prepared for launch to Mars at every twenty-six-month opportunity—two craft were launched right on schedule in the December 1998–January 1999 window (and unfortunately failed as they arrived at Mars)—and plans are progressing on new mini-missions to Europa, Pluto, and several asteroids and comets. The emphasis is on Mars, and the planned program includes orbiters, landers, penetrators, rovers, sample return vehicles, an experiment to demonstrate propellant production from the Martian atmosphere, and even perhaps a Mars mini-airplane. All of this activity marks what promises to be a vigorous new era of solar system exploration. It was heartening to observe that public interest built again commensurate with the increased mission frequency: there were hundreds of millions of contacts with the NASA web sites during the 1997 Pathfinder/Sojourner operations on the surface of Mars.

NASA's challenging robot spacecraft have always experienced in-flight problems that have to be solved by their dedicated operators. Such problems should not come as a shock. Although NASA needs to strive always to eliminate faults and design flaws, one hopes that an occasional failure will not cause the "faster, better, cheaper" approach to waiver. "Big" spacecraft should be used only when really required to achieve primary mission objectives—small failures are a lot easier to shake off than multi-billion-dollar ones.

The indefatigable Bronx-raised Lou Friedman, executive director of the Planetary Society, is a great fan and staunch supporter of the inde-fatigable Bronx-born Dan Goldin. If we indeed do achieve another golden era of planetary exploration, Lou proposes that it be known as the "Goldin" era (pun fully intended). Personally, I think Dan Goldin was sometimes unnecessarily rough in replacing NASA managers, but in 1992 he immediately recognized the austere budget climate the agency was facing and instituted strong measures to increase the effi-ciency within NASA, doing more with fewer people and dollars. Con-sequently, NASA was still able to pursue a vast array of valuable proj-ects in spite of an ever slipping budget. Of course, OMB loved this, and as a result the Republican-appointed Goldin was continued as NASA administrator by the newly elected Democratic president Bill Clinton. A pretty rare carry over. Dan Goldin may have his detractors, especially among NASA managers who have been "fired," but there is no deny-ing his dedication and tireless effort in advancing the space age. How-ever, he must face the unenviable task of completing an enormously complex multi-billion-dollar International Space Station (ISS) project without squeezing out space science, as happened with the shuttle. That objective is virtually impossible unless Dan can somehow reverse the steady year-after-year decline that the NASA budget experienced under the first seven years of the Clinton administration.

Looking to the future, in spite of the budget obstacles, I am encour-aged by the fact that planetary exploration has always attracted tal-ented and highly motivated people in all supporting professions and at all skill levels. Exploration provides a powerful motivation. As long as that spirit is alive, and I believe it is innate in the human species, we should continue to see a steady string of bold missions to further study our solar system and the universe beyond. I like very much the follow-ing quote found by the space historian William Burrows from a 1936 science fiction movie: "But for Man, no rest and no ending. He must go on, conquest beyond conquest. First this little planet with its winds and ways, and then all the laws of mind and matter that restrain him. Then the planets around him, and at last out across immensity to the stars. And when he has conquered all the deeps of space and all the mysteries of time, still he will be beginning."[6]

Lessons Learned

Now comes a difficult part for me to write, if I am to be fair. Hindsight is much easier than foresight and on-the-spot decision making. Just ask the tormented few who gave the OK for the cold-morning launch of Challenger in January 1986.

NASA's problems have never been attributable to a lack of talented and dedicated people. Its charter for exploration and advancing technology has always attracted the best engineers, scientists, and administrators. High-paid executives such as Tom Paine, Jim Fletcher, Jim Beggs, and Dan Goldin have left top positions in the aerospace industry to serve as its ridiculously underpaid administrator. The biggest problem internal to NASA is the decades-old dilemma that has confronted all of NASA's administrators: keeping manned space flight moving forward. Every administrator has concluded that a program keeping humans in space is essential to public support, but that has led to very damaging decisions for Space Science. I have no magic solution to that dilemma. Peripheral to that big issue, though, and related to manned spaceflight and its all-too-frequently negative impact on Planetary Programs, I do see two lessons NASA should have learned.

LESSON ONE
Do Not Let Politics Dictate Technical Decisions

Easier said than done, but nevertheless important. Jim Fletcher should not have let OMB trap him into justifying the space shuttle purely on economics, promising that it would substantially reduce the cost of putting all kinds of payloads into space. That forced many missions onto the shuttle that had no reason to be there. We wound up with all our eggs in the shuttle basket and were brought to a halt by the Challenger failure.

Did we learn from that? Regrettably, no. In trying with great effort and dedication to sell the next planned step in manned spaceflight, an orbiting space station, NASA administrator Jim Beggs started emphasizing its great payoff in developing new and marvelous zero-gravity manufacturing techniques in space. That should eventually come about, but probably slower than hoped and hyped. And if you really want zero gravity in orbit you do it with unmanned vehicles. Humans moving around in a space station, machinery operating, and station attitude and altitude adjustments cause accelerations and vibrations that destroy zero-gravity conditions. Even if one settles for micro-gravity rather than zero-gravity conditions, it is almost impossible to achieve the desired one-millionth of a g. The International Space Station will no doubt be a valuable research facility and produce many practical outputs and spinoffs, but it should be recognized for what it primarily is—a necessary step to developing long-term life-support systems and techniques for future missions sending humans to Mars. Wernher von Braun outlined in the *Collier's* series of 1952–54 the logical sequence for advancing the human presence in the solar system: first developing a reusable shuttle, followed by an orbiting space station, and then a mission sending humans to land on Mars. Planning group after planning group have validated the logic of that sequence and the essential role for a space station. Setting foot on Mars has been a dream and objective for the human race for more than one hundred years, so why deny it? Sell the space station for what it primarily is: a step to Mars.

As prospects for manufacturing breakthroughs in space appeared less imminent, the fledgling space station project was facing cancellation in Congress. What saved it was bringing the Russians aboard as a full partner. The reconstituted ISS became a symbol of international cooperation between the world's superpowers. Perfectly correct and very inspiring, but let us be honest and keep it headed for its true objective. For example, a major question that needs to be answered before humans head for Mars is whether the human body can stand a voyage of more than one year in a zero-gravity (or micro-gravity) environment. If not, the Mars spacecraft will have to spin and be of large enough diameter to avoid adverse Coriolis effects that would disorient the Mars-bound crew, and this would add substantially to the weight required. The major goal of research on the ISS should be to provide the necessary data to make that critical decision between zero gravity versus a spinner.

Lou Friedman, cofounder and executive director of the Planetary Society, said it well in the February 1998 issue of *Space News:* "The goal of America's human spaceflight program—openly stated or not—is the exploration of Mars." Wernher von Braun saw that clearly. So did Carl Sagan. Let us not be ashamed or too cautious to admit it. I liked the following editorial comment in the *Washington Post* the day after the first module of the International Space Station was launched: "There's never been an entirely logical justification for the [manned] space program beyond the fundamental one—that space seemed the obvious next destination for a species of explorers. The space station, in that sense, is the obvious next step to prepare for the step after. Whether that turns out to be interplanetary travel, moon colonization or something we can't today foretell isn't clear. But that, in a way, is the point."[7]

Did NASA do a better job of separating politics from projects in the past? I think so. Take the example of nuclear rocket propulsion in the 1950s to 1970s. A rocket engine's efficiency (specific impulse) is proportional to the square root of the combustion temperature divided by the molecular weight of its exhaust gases. A nuclear reactor can produce very high temperatures, and it can heat hydrogen gas, which has the lowest molecular weight of all the elements. So a hydrogen-fueled nuclear rocket can be a super performer. Technology development was started in the 1950s by the AEC under the code names of Rover, Kiwi, and, later, NERVA. When NASA was created it also contributed financial support. Aerojet Corporation designed and built thrust chambers for the program, and Rocketdyne started development of a turbopump for liquid hydrogen. By far the strongest supporter of the project in Congress was Sen. Clinton Anderson of New Mexico, the home state of the AEC's Los Alamos Laboratory, where the reactor work was based.

Senator Anderson was chairman of the Senate Aeronautical and Space Sciences Committee, which had to pass on NASA's budget authorization each year. NASA paid a lot of attention to Senator Anderson and he to NASA, especially on how NASA planned to use the nuclear rocket engine. At NASA Headquarters in 1968 I was appointed chairman of a Nuclear Rocket Missions Study Panel to identify where NASA was going to use the NERVA engine. We found that the reactor was inherently so heavy that it only made sense to use it in a launch vehicle's upper stage with a huge tank of liquid hydrogen propelled by a low-thrust engine—in other words, an engine with a thrust-to-weight ratio of far less than one. (With this F/W of less than one, it could not be used in a booster stage to rise off the surface.) Even with the small (seventy-five thousand pounds of thrust) NERVA engine, the upper stage had to be so large that it would take NASA's largest booster, the immense and expensive Saturn 5, to get it off the ground. The only mission that required that capability would be one to carry humans to Mars, and Congress had already turned thumbs down on that for the near future. So the study panel recommended shutting down the NERVA program. NASA management agreed and deleted it from the next year's budget request.

Senator Anderson hit the roof. He had been supporting the administration's proposed development of a supersonic transport (SST), but got so mad at the plan to shutdown NERVA that he switched his vote and talked one his committee members, Sen. Margaret Chase Smith of Maine, into also voting nay. Those two votes were the margin that killed the SST. His target at NASA was the Grand Tour proposal, reasoning that without the multiple gravity assists of the Grand Tour NASA would have to use NERVA to achieve fast trips to the outer planets. Although John Naugle and Jim Fletcher did indeed try to placate Anderson, I am proud to say that NASA stuck by what was technically correct and finally managed to terminate its effort on NERVA. (Although more than $500 million had been spent, the technology will be useful in the future, and the hydrogen turbopump work was applied directly to the design of the hydrogen pump for the Saturn 5's J-2 upper-stage engines.) Grand Tour, of course, did survive (under the name of Voyager).

LESSON TWO

Speak Straight with Congress

In the 1970s, when we were in deep technical troubles with the development of the Viking Mars Landers, we bent over backward to keep congressional committees informed. Jim Gehrig, Senator Anderson's able right-hand man and chief of staff, was especially interested. I went every two weeks, or however often he wanted, to brief him and fellow staffers Glen Wilson and Craig Vorhees on all of our difficult problems, what we were doing about them, and what the impact might be on project costs. Over on the House side, twice we flew a group of committee members and staff down to NASA Langley to get an all-day detailed briefing on the project status. Throughout the life of the Viking Project we kept the committees informed of how our budget reserves were holding up and whether we were facing an overrun in the budget for future years. Largely because of this frankness and complete honesty, I believe, we had complete support in Congress throughout the years of technical struggle.

Contrast that with the early turbulent history of the space station project. Too many NASA managers gave conflicting testimony to the committees. Forecasts of schedule and costs were blatantly optimistic. Dan Goldin was embarrassed frequently by apparently not being given accurate and up-to-date information. It has reached the point where before every hearing of the House Authorization Subcommittee the chairman requires all NASA members present to stand up, raise their right hands, and swear to tell the truth and the whole truth to the committee. I cringe every time I witness this. Dan Goldin handles it well, but I am sure he resents the implication that he is deliberately lying. This difficult period appears now to be over, and congressional committee members express the highest respect for Dan Goldin. Thank goodness. All NASA employees, and we old retirees, want to continue to be proud of our agency and its reputation for integrity.

No more of my sanctimonious preaching on lessons learned. Here's to the marvels yet to be discovered in our solar system and the galaxy-studded universe, and may our children and grandchildren exult in their walks upon Mars.

APPPENDIX

NASA PLANETARY PROBES, 1970–1980

Mission / Launch Date	Comments
Mariner 8 6 May 1971	Mariner 8 was (along with Mariner 9) part of the Mariner Mars 71 project. It was intended to go into Mars orbit and return images and data. Mariner 8 was launched on an Atlas-Centaur SLV-3C booster (AC-24). The main Centaur engine was ignited 265 seconds after launch, but the upper stage began to oscillate in pitch and tumbled out of control. The Centaur and spacecraft payload separated and reentered the Earth's atmosphere.
Mariner 9 30 May 1971	This spacecraft was a sister spacecraft to Mariner 8. Mariner 9 combined mission objectives of both Mariner 8 (mapping 70 percent of the Martian surface) and Mariner 9 (a study of temporal changes in the Martian atmosphere and on the Martian surface).
Pioneer 10 2 March 1972	This mission was the first to be sent through the Asteroid Belt to the outer solar system and the first to investigate the planet Jupiter, after which it followed an escape trajectory from the solar system, heading outward from the Sun at 2.6 AU/year and downstream through the heliomagnetosphere

Mission / Launch Date	Comments

toward the tail region and interstellar space. The spacecraft achieved its closest approach to Jupiter on 3 December 1973 when it reached approximately 2.8 Jovian radii (about 200,000 kilometers). Routine tracking and project data processing operations were terminated on 31 March 1997 for budget reasons. Occasional tracking continues under support of the Lunar Prospector Project at NASA Ames Research Center with retrieval of energetic particle and radio science data.

Pioneer 11
5 April 1973

This was the second mission to investigate Jupiter and the outer solar system and the first to explore the planet Saturn and its rings. During its closest approach, 4 December 1974, Pioneer 11 passed to within 34,000 kilometers of Jupiter's cloud tops. It passed by Saturn on 1 September 1979 at a distance of 21,000 kilometers from Saturn's cloud tops. Science operations and daily telemetry ceased on 30 September 1995 when the RTG power level was insufficient to operate any experiments.

Mariner 10
3 November 1973

Mariner 10 was the seventh successful launch in the Mariner series and the first spacecraft to use the gravitational pull of one planet (Venus) to reach another (Mercury). The spacecraft flew by Mercury three times in a heliocentric orbit and returned images and data on the planet. The spacecraft passed Venus on 5 February 1974, at a distance of 4,200 kilometers. It crossed the orbit of Mercury on 29 March 1974, at 2,046 UT, at a distance of about 704 kilometers from the surface. A second encounter with Mercury occurred on 21 September 1974 at an altitude of about 47,000 kilometers. A third and last Mercury encounter at an altitude of 327 kilometers occurred on 16 March 1975. Engineering tests were continued until 24 March 1975, when the supply of attitude-control gas was depleted and the mission was terminated.

Helios 1
12 October 1974

This spacecraft was one of a pair of deep-space probes developed by the Federal Republic of Germany in cooperation with NASA. NASA sup-

Mission / Launch Date	Comments
	plied the Titan-Centaur launch vehicle. The purpose of the mission was to make pioneering measurements of the interplanetary medium from the vicinity of Earth's orbit to .3 AU. The payload consisted of a fluxgate magnetometer; electric and magnetic wave experiments; charged-particle experiments, which covered various energy ranges starting with solar wind thermal energies; a zodiacal-light experiment; and a micrometeoroid experiment.
Viking 1 20 August 1975	This spacecraft consisted of an orbiter and a lander. It went into orbit around Mars on 19 June 1976. Its Lander touched down on 20 July 1976 on the western slopes of Chryse Planitia (Golden Plains). The Lander had experiments to search for Martian micro-organisms, provided detailed color panoramic views of the Martian terrain, and also monitored the Martian weather. The Orbiter mapped the planet's surface, acquiring over 52,000 images. The Viking 1 orbiter was deactivated on 7 August 1980 when it ran out of attitude-control propellant. Viking 1 Lander was accidentally shut down on 13 November 1982, and communication was never regained.
Viking 2 9 September 1975	This spacecraft was a sister to Viking 1 and consisted of an orbiter and a lander. Viking 2 landed on Mars on 3 September 1976 in a rocky area of Utopia Planitia at 48 degrees north latitude. The Viking 2 Orbiter ran out of attitude-control propellant on 25 July 1978, and the Viking 2 Lander was shut down on 12 April 1980.
Helios 2 15 January 1976	This spacecraft was the second of a pair of deep-space probes developed by the Federal Republic of Germany in a cooperative program with NASA. Helios 2 was launched on a Titan-Centaur rocket provided by NASA. The purpose of the mission was to make pioneering measurements of the interplanetary medium from the vicinity of Earth's orbit to .29 AU.
Voyager 1 5 September 1977	Once every 176 years both the Earth and all the giant planets of the solar system gather on one side

Mission / Launch Date	Comments
	of the Sun. This geometric lineup made possible close-up observation of all the planets in the outer solar system (with the exception of Pluto) in a single flight, the Grand Tour. NASA launched two of these: Voyager 2 lifted off on 20 August 1977, and Voyager 1 entered space on a faster trajectory on 5 September 1977. Both spacecraft were delivered to space aboard Titan Centaur expendable rockets. In February 1979 Voyager 1 entered the Jovian system and then sped on to Saturn and its large satellite Titan.
Voyager 2 20 August 1977	Voyager 2 explored Jupiter's moons and then traveled on to fly by Saturn in July 1981. With the successful achievement of all its objectives at Jupiter and Saturn, additional flybys by Voyager 2 of the two outermost giant planets, Uranus and Neptune, proved possible. In January 1986 Voyager 2 encountered Uranus, and in 1989 it encountered Neptune. Eventually, between them, Voyager 1 and Voyager 2 explored all the giant outer planets, forty-eight of their moons, and the unique systems of rings and magnetic fields those planets possess.
Pioneer Venus 1 20 May 1978	The Pioneer Venus Orbiter was inserted into an elliptical orbit around Venus on 4 December 1978. The spacecraft was in a 24-hour orbit with an apoapsis of 66,900 kilometers. After its radar altimeter mapped the planet's topography, in May 1992 Pioneer Venus began the final phase of its mission, in which the periapsis was held between 150 and 250 kilometers until the fuel ran out and atmospheric entry destroyed the spacecraft the following August.
Pioneer Venus 2 8 August 1978	The Pioneer Venus Multiprobe consisted of a bus that carried one large and three small atmospheric probes. All four probes entered the Venus atmosphere on 9 December, followed by the bus. The Pioneer Venus large probe was equipped with seven science experiments, contained within a sealed spherical pressure vessel. The probe radioed data until destroyed by pressure and temperature.

NOTES

CHAPTER 1. On the Shoulders of Giants

1. Robert H. Goddard, "A Method of Reaching Extreme Altitudes," *Smithsonian Miscellaneous Collections* 71, no. 2, Washington, D.C., 1919.
2. Konstantin Tsiolkovsky, "Exploring Space with Reactive Devices," *Scientific Review,* 1926.
3. Hermann Oberth, *Die Rakete zu den Planetenraumen* (Rockets in Planetary Space) (Munich: Verlag von R. Oldenbourg), 1923.
4. These design details were not reported at the time as Goddard had become very reclusive and secretive about his work. In the early 1940s, he volunteered to work with the navy in Annapolis, Maryland, to develop a liquid-propellant jet-assist takeoff rocket. Young, crew-cut navy lieutenant Bob Truax was working nearby trying to develop a similar JATO but using noncryogenic "storable" propellants. Bob later related that he made repeated efforts to share test results and get guidance from Goddard but was always rebuffed.
5. Michael J. Neufeld, *The Rocket and the Reich: Peenemünde and the Coming of the Ballistic Missile Era* (New York: Free Press, 1995), 55.
6. Although von Braun was an inspiration and leader in creating new launch vehicles in the United States, rocket-engine development quickly departed from his V-2 design on two independent paths. The Russians milked the knowledge of the Germans captured at Peenemünde but then

went their own way, using multiple, small thrust chambers with each turbopump to minimize problems with the destructive combustion instability frequently encountered in larger thrust chambers. In the United States, Rocketdyne, following delivery of V-2–descendant engines for von Braun's Redstone vehicle, advanced to much lighter thrust chambers made up of bundles of thin-walled tubes and then focused on simplifying the entire propulsion system. The V-2 engine's eighteen premix chambers, which required a maze of intertwined plumbing leading to the thrust chamber, were replaced with a single flat-plate injector. Then Rocketdyne engineers eliminated the German's multivalved staged start sequence (which some of the Peenemünde engineers, such as the rather arrogant Walter "Papa" Riedel, insisted was essential) in favor of a rapid start that proved to be very smooth and not prone to combustion instability. Then they eliminated the pneumatic and hydraulic systems and reduced the electrical system to little more than two wires. After verification in Rocketdyne's X-1 experimental engine, this simplified system went into production as the H-1 engine for the Saturn 1 launch vehicle.

7. There were other "sensitive" reasons for the first U.S. orbiter to be purely scientific. Eisenhower wanted no protests from the Soviet Union about violating their space, as the high-altitude U-2 reconnaissance planes had done. The Soviets could not very well protest a scientific civilian orbiter launched as a U.S. contribution to the International Geophysical Year. Meanwhile, a highly classified WS-117L satellite reconnaissance program, its early components called Corona and Discoverer, could proceed under cover.

8. Charlie Bossart was an excellent engineer, but Convair apparently did not think he could manage a project as large as Atlas. As the program accelerated, we at Rocketdyne were soon dealing with "smoother" managers such as Charlie Ames, and then Convair brought in a dynamic thirty-four-year-old red-headed "whiz kid" named James Dempsey to head up the new company division created around Atlas. Bossart, however, provided the initial inspiration and breakthrough design.

9. A fifth Pioneer was built by NASA's Goddard Space Flight Center and successfully launched into a one-AU orbit around the Sun on 11 March 1960.

10. One evening in the early 1970s, while having a pleasant dinner conversation with Esther Goddard, widow of Robert Goddard, I mentioned that in my early years as a rocket engineer I had found very dependable and useful the calculations of Frank Malina. That lovely lady turned on me and hissed, "Don't you ever mention that horrible man's name to me again!" She then told me of the prolonged court battle between Malina

and Robert Goddard over rocket patents. Too bad that these two rocket pioneers had to wind up as adversaries.

11. I hated to leave Rocketdyne. In my eleven years there I believe I was the first member of management to leave. I made good friends for life, and I would not trade those years for anything. What greater thrill for a young space enthusiast than to help design great rocket engines and then stand recklessly nearby (without sandbags or even ear protectors) to watch and feel the immense power and fiery exhaust of all that suddenly released energy.

12. At Aeronutronic we were able to simulate Gold Dust by blowing air up through Cab-O-Sil, a commercial thickener composed of rough-surfaced microscopic particles of silica. A barrel of the fluffed Cab-O-Sil looked like solid white sand, but if one dropped a soccer ball into the barrel, the ball would bounce off the bottom of the barrel, passing through the "sand" as if it were fog.

13. The AIAA was created in 1963 by the merger of the American Rocket Society and the venerable Institute of Aeronautical Science.

14. The heroic crewman presented NASA management with a dilemma. He had broken strict safety rules and had endangered his life, but he had saved a valuable spacecraft and launch vehicle. Should he be fired or rewarded? NASA chose the latter, but did it quietly.

CHAPTER 2. 1971: Mariner 9 Mars Orbit

1. The tragic fire during a ground test of Apollo 1 on 27 January 1967, which took the lives of astronauts Gus Grissom, Roger Chaffee, and Edward White, precipitated the dissolution of the outstanding team of Webb and his very able deputy, Robert Seamans. After the fire Webb apparently put some of the blame on Seamans, who felt compelled to resign. Similarly, Lyndon Johnson put the responsibility on Webb and forced him to resign. It was really an injustice that Seamans and, especially, Webb were not at NASA to take their due credit for the triumph of Apollo 11's historic landing on the Moon on 20 July 1969.

2. The name and charter of Code SL changed from time to time. During the 1960s it was Lunar and Planetary Programs and directed the unmanned lunar projects Ranger, Surveyor, and Lunar Orbiter. During the Apollo landings (1969 to 1972) lunar science was placed under Apollo management and SL was known as just Planetary Programs. Following the final Apollo 17 mission, responsibility for the Apollo lunar samples and for the continuing operation of the Apollo Lunar Surface Experiments

Package (ALSEP) instruments left on the Moon returned to SL, which regained its former name of Lunar and Planetary Programs. Eventually the title Lunar and Planetary would be broadened to Solar System Exploration.

3. Arnold Frutkin and I did not always agree at work, but we became friends and partners in a cruising sailboat. He was a gentleman, a fine sailor, and meticulous in just about everything, which made him an excellent boating partner.

4. Bruce C. Murray, *Journey into Space* (New York: W. W. Norton, 1989), 62.

5. Carl Sagan, *Pale Blue Dot: A Vision of the Human Future in Space* (New York: Random House, 1994), 232.

6. *Mars as Viewed by Mariner 9*, NASA SP-329 (Washington, D.C.: GPO, 1974), 185.

7. Oran Nicks, *Far Travelers: The Exploring Machines*, NASA SP-480 (Washington, D.C.: GPO, 1985), 171.

8. Clayton R. Koppes, *JPL and the American Space Program* (New Haven, Conn.: Yale University Press, 1982), 220.

CHAPTER 3. 1972 and 1973: Pioneer 10 and 11 Jupiter and Saturn Flybys

1. NASA History Office, *NASA Pocket Statistics*, 1995 ed.

CHAPTER 4. 1974 and 1976: Helios 1 and 2 Solar Probes

1. Herbert Porsche, ed., *10 Years HELIOS* (Munich: Wenschow Franzis Druck GmbH, 1984), 9.

CHAPTER 5. 1973: Mariner 10 Venus and Mercury Flybys

1. *Terrestrial* is a term astronomers use in reference to the rocky planets Mercury, Venus, Earth, and Mars, as opposed to the giant gas balls of the outer planets, Jupiter, Saturn, Uranus, and Neptune. Based on current knowledge, and assuming it is an accreted planet rather than an "escaped" satellite, Pluto would have to be classified with the terrestrial group.

2. Nicks, *Far Travelers*, 238.

CHAPTER 6. 1975: Viking 1 and 2 Mars Orbits and Landings

1. Christopher P. McKay and Wanda L. Davis, "Planets and the Origin of Life," in *Encyclopedia of the Solar System,* ed. Paul R. Weissman, Lucy-Ann McFadden, and Torrence V. Johnson, 911 (San Diego: Academic Press, 1999).
2. Percival Lowell, *Mars* (Boston: Houghton Mifflin, 1895).
3. Dr. Fred L. Whipple, "Is There Life on Mars?" *Collier's,* 30 April 1954, 21.
4. Yervant Terzian and Elizabeth Bilson, eds., *Carl Sagan's Universe* (Cambridge: Cambridge University Press, 1997), 40.
5. In 1963 the Office of Space Science was renamed the Office of Space Science and Applications, which reverted in 1971 back to the Office of Space Science.
6. This was not Oran Nicks's last involvement with Lunar and Planetary Programs, whose creation he had led so successfully. He later served as deputy director of the Langley Research Center during the peak years of the Viking Project. Sadly, he died in an accident in 1998 doing one of the things he loved best, flying his sailplane. I was privileged to have worked both with him and for him, and to have been among his friends for more than forty years. His lovely and talented wife, Phyllis, and their children can be proud of all that Oran accomplished in a most productive and pioneering professional career.
7. Robert C. Seamans Jr., *Aiming at Targets,* NASA SP-4106 (Washington, D.C.: GPO, 1996), 82.
8. The minimum-energy path to another planet is along a Hohmann ellipse around the Sun, which for Mars means leaving Earth at the periapsis of the ellipse and arriving at Mars at the apoapsis. Opportunities for this Hohmann ellipse launch to Mars occur every twenty-six months and to Venus every nineteen months.
9. Richard J. Blott, UK Defence Evaluation and Research Agency, in remarks to a small-satellite conference in Juan-les-Pins, France, on 16 September 1998, as reported in the 21–27 September 1998 issue of *Space News.*
10. Today there is more concern about global warming due to the CO_2 greenhouse effect (revealed by the fiercely hot planet Venus) than about any imminent ice age. However, the Mars dust/temperature observations were effectively used by Carl Sagan and associates to calculate that a nuclear war between the United States and the Soviet Union would stir up enough dust in the atmosphere to create a "nuclear winter" severe enough to kill all life on Earth.

11. Bruce Murray succeeded William Pickering as director of JPL in April 1976, just in time for the Viking landings and to help get the Voyager Project through some crisis years. He had for years offered unsolicited advice to NASA and various congressional committees on how the space program should be run, and now he had a chance to influence matters from within NASA. Yet during his six years as director of JPL he was frustrated by the lack of NASA funding for space science and an inability to sell his proposed new missions (including a fascinating solar sailing visit to a comet). Nevertheless, one of his successful and lasting creations during that period was the Planetary Society, which he cofounded with Carl Sagan and Louis Friedman in 1980. Since Carl's untimely death in December 1996, Bruce has served as president of the very active 100,000-member society.

CHAPTER 7. 1977: Voyager 1 and 2 Flybys of the Outer Planets

1. *National Geographic* 138, no. 2 (August 1970): 193.
2. David Swift, *Voyager Tales: Personal Views of the Grand Tour* (Reston, Va.: American Institute of Aeronautics and Astronautics, 1997), 81.
3. William E. Burrows, *Exploring Space* (New York: Random House, 1990), 290.
4. We learned later that just before Voyager's encounter three scientists, Stanton Peale, Patrick Cassen, and Ray Thomas Reynolds, had submitted a paper to "Science" predicting that tidal heating could indeed induce volcanism on Io.
5. Burrows, *Exploring Space,* 299.
6. Ibid.
7. Ibid.
8. Ronald A. Schorn, *Planetary Astronomy: From Ancient Times to the Third Millennium* (College Station: Texas A&M University Press, 1998), 283.
9. Terzian and Bilson, *Carl Sagan's Universe,* 152.
10. Sagan, *Pale Blue Dot,* 4.
11. Swift, *Voyager Tales,* 188.
12. Ibid., 131.

CHAPTER 8. 1978: Pioneer Venus 1 and 2 Orbit and Probes

1. Charles Hall, interview with author, Los Altos, California, 22 May 1997.
2. Richard O. Fimmel, Lawrence Colin, and Eric Burgess, *Pioneer Venus,* NASA SP-461 (Washington, D.C.: GPO, 1983), 99–100.

Epilogue

1. Remarks made at the San Francisco meeting of the American Association for the Advancement of Science in February 1974, as reported in *Astronomy* 2, no. 6 (June 1974).
2. This starvation of eleven years without the launch of a new mission was made less apparent to the public by the stretched-out planetary encounters of the two Grand-Touring Voyager '77 spacecraft: Jupiter in 1979, Saturn in 1980 and 1981, Uranus in 1986, and Neptune in 1989.
3. Several times I thought I had negotiated a deal that would give JPL overall systems responsibility for an SEP-powered solar system exploration mission, with ion-thruster technical support from LeRC. Each time MSFC blocked progress by pushing for a general-purpose SEP stage under their responsibility.
4. Technically, the United States could claim the first comet encounter. Through the trajectory ingenuity of Bob Farquhar of GSFC, a small interplanetary Explorer spacecraft (the International Sun-Earth Explorer, or ISEE, launched in 1978) was diverted via five very close passes by the Moon to pass through the tail of Comet Giacobini-Zinner on 11 September 1985. However, the spacecraft, renamed ICE, carried no comet-specific instruments.
5. I rather regret that I was no longer at NASA Headquarters during Bruce Murray's vigorous campaign to sell a Comet Halley mission. Bruce and I communicated well, far better than I did with Bill Pickering. For example, when Bruce on sabbatical leave from Caltech first learned in 1975 of his selection to direct JPL he invited me to the house he and Suzanne had rented in La Jolla, California. We spent a day sharing views on JPL management and relations with NASA Headquarters. For the Comet Halley rendezvous mission, although I would have tried to swing Bruce over from solar sailing to less "purple" but technically proven and more saleable solar electric propulsion, I would have enjoyed working together with him, like Don Quixote in his pursuit of an impossible dream, fighting against adversity to sell a Halley mission.
6. *Things to Come*, a film released in 1936 based on H. G. Wells's novel *The Shape of Things to Come*.
7. *Washington Post,* editorial, "Construction in Space," 21 November 1998, p. A20.

REFERENCES

Published Sources

Aerojet History Group. *Aerojet: The Creative Company.* Los Angeles: Stuart F. Cooper, 1995.
"Aliens in a Slushy Sea? Evidence Mounts for Life on a Wet Jovian Moon." *Time,* 16 March 1998.
American Geophysical Union. "Pioneer Venus." *Journal of Geophysical Research* 85, no. A13 (30 December 1980): 7573–8337.
———. "Scientific Results of the Viking Project." *Journal of Geophysical Research* 82, no. 28 (30 September 1977): 3959–4681.
Biemann, Hans-Peter. *The Vikings of '76.* Westford, Mass.: Murray Printing, 1977.
Bradbury, Ray, Arthur C. Clarke, Bruce Murray, Carl Sagan, and Walter Sullivan. *Mars and the Mind of Man.* New York: Harper & Row, 1973.
Burrows, William E. *Exploring Space.* New York: Random House, 1990.
———. *This New Ocean: The Story of the First Space Age.* New York: Random House, 1998.
Clarke, Arthur C., ed. *The Coming of the Space Age.* New York: Meredith Press, 1967.
Cooper, Henry S. F., Jr. *Imaging Saturn: The Voyager Flights to Saturn.* New York: Holt, Rinehart, and Winston, 1982.

Corliss, William R. *Planetary Exploration*. NASA Space in the Seventies Series. Washington, D.C.: GPO, 1971.

———. *The Viking Mission to Mars*. NASA SP-334. Washington, D.C.: GPO, 1975.

Dunne, James A., and Eric Burgess. *The Voyage of Mariner 10: Mission to Venus and Mercury*. NASA SP-424. Washington, D.C.: GPO, 1978.

Ezell, Edward Clinton, and Linda Neuman Ezell. *On Mars: Exploration of the Red Planet 1958–1978*. NASA SP-4212. Washington, D.C.: GPO, 1984.

Fimmel, Richard O., James Van Allen, and Eric Burgess. *Pioneer: First to Jupiter, Saturn, and Beyond*. NASA SP-446. Washington, D.C.: GPO, 1980.

Fimmel, Richard O., Lawrence Colin, and Eric Burgess. *Pioneer Venus*. NASA SP-461. Washington, D.C.: GPO, 1983.

Fimmel, Richard O., William Swindell, and Eric Burgess. *Pioneer Odyssey: Encounter with a Giant*. NASA SP-349. Washington, D.C.: GPO, 1974.

Friedman, Louis. "Rendezvous with Infinity." *Space News* 9, no. 7 (16–22 February 1998).

Friedman, Louis, and Robert Kraemer. "Planetary Exploration Missions." In *Encyclopedia of the Solar System*, ed. Paul R. Weissman, Lucy-Ann McFadden, and Torrence V. Johnson, 923–39. San Diego: Academic Press, 1999.

Goddard, Robert H. "A Method of Reaching Extreme Altitudes." *Smithsonian Miscellaneous Collections* 71, no. 2. Washington, D.C., 1919.

———. "Liquid-propellant Rocket Development." *Smithsonian Miscellaneous Collections* 95, no. 3, Smithsonian Institution Press, 1936.

Gore, Rick. "Uranus: Voyager Visits a Dark Planet." *National Geographic*, August 1986.

Grey, Jerry. "Commentary." *Aerospace America* 35, no. 12 (December 1997).

Hall, R. Cargill. *Project Ranger: A Chronology*. JPL/HR-2, 1971.

Hartman, William K., and Odell Raper. *The New Mars: The Discoveries of Mariner 9*. NASA SP-337. Washington, D.C.: GPO, 1974.

Heppenheimer, T. A. *Countdown: A History of Space Flight*. New York: John Wiley & Sons, 1997.

Hoban, Frank. "Gaining Steps with Low-Cost Systems." *Space News*, 14–20 April 1997.

Holdes, William G., and William O. Siuru Jr. *Skylab: Pioneer Space Station*. New York: Rand McNally, 1974.

Hotz, Robert. "Voyager's New Worlds." Editorial, *Aviation Week & Space Technology*, 9 April 1979.

Jaroff, Leonard. "The Last Time We Saw Mars." *Time*, 14 July 1997, 35.

————. "Still Ticking: A Quarter Century after Its Launch, Pioneer 10 Is Alive and Calling Home." *Time*, 4 November 1996, 80.

Kohlhase, Charles. *The Voyager Neptune Travel Guide*. JPL Publication 89–24. Pasadena, Calif.: JPL, 1989.

Koppes, Clayton R. *JPL and the American Space Program*. New Haven, Conn.: Yale University Press, 1982.

Kraemer, Robert S. "Mars: Tomorrow's Earth?" *Cincinnati Enquirer*, 30 May 1976, pp. 14–19.

————. "Out of Step with Reality? Traipsing among the Planets May Help Man Solve His Problems." *Empire Magazine, Denver Post*, 6 June 1976, pp. 10–13.

————. "Planets Reflect Earth's Future." *Today*, 30 May 1976, pp. 1G–3G.

————. "The Planets: Reflections of Future Earth." *Nevadan*, 9 May 1976, pp. 3–5.

————. "Space Study Is Down-to-Earth: Venus Sends Us a Warning." *Times Weekend*, 15 May 1976, pp. 2A–4A.

Lambright, W. Henry. *Powering Apollo: James E. Webb of NASA*. Baltimore: Johns Hopkins University Press, 1995.

Lasher, Larry. "Pioneer 10: A Model of Success." *Space News*, 7–13 April 1997, 15.

Launius, Roger D. *NASA: The History of the U.S. Civil Space Program*. Malabar, Fla.: Krieger Publishing, 1994.

Ley, Willy. *Rockets, Missiles, and Space Travel*. New York: Viking Press, 1951.

Logsdon, John M., ed. *Exploring the Unknown: Selected Documents in the History of the U.S. Civil Space Program*. Vol. 1, *Organizing for Exploration*. NASA SP-4218. Washington, D.C.: GPO, 1995.

Lowell, Percival. *Mars*. Boston: Houghton Mifflin, 1895.

Mariner-Mars 1964: Final Project Report. NASA SP-139. Washington, D.C.: GPO, 1967.

Mariner-Mars 1969: A Preliminary Report. NASA SP-225. Washington, D.C.: GPO, 1969.

Mars as Viewed by Mariner 9. NASA SP-329. Washington, D.C.: GPO, 1974.

McCurdy, Howard E. *Inside NASA: High Technology and Organizational Change in the U.S. Space Program*. Baltimore: Johns Hopkins University Press, 1993.

————. *Space and the American Imagination*. Washington, D.C.: Smithsonian Institution Press, 1997.

Miller, Ron, and William K. Hartmann. *The Grand Tour: A Traveler's Guide to the Solar System*. 1981. Rev. ed., New York: Workman, 1993.

Morrison, David. *Voyages to Saturn*. NASA SP-451. Washington, D.C.: GPO, 1982.

Murray, Bruce C. *Journey into Space*. New York: W. W. Norton, 1989.

NASA History Office. *NASA Pocket Statistics*. 1995 edition.

Naugle, John E. *First Among Equals: The Selection of NASA Space Science Experiments*. NASA SP-4215. Washington, D.C.: GPO, 1991.

Neufeld, Michael J. *The Rocket and the Reich: Peenemünde and the Coming of the Ballistic Missile Era*. New York: Free Press, 1995.

Newell, Homer E. *Beyond the Atmosphere: Early Years of Space Science*. NASA SP-4211. Washington, D.C.: GPO, 1980.

Nicks, Oran W. *Far Travelers: The Exploring Machines*. NASA SP-480. Washington, D.C.: GPO, 1985.

Ordway, Frederick I., III, and Mitchell R. Sharpe, *The Rocket Team*. New York: Thomas Y. Crowell, 1979.

The Pioneer Mission to Jupiter. NASA SP-268. Washington, D.C.: GPO, 1971.

The Planetary Society. "Carl Sagan: A Tribute." *Planetary Report* 17, no. 3 (May/June 1997): 4–22.

Porsche, Herbert, ed. *10 Years HELIOS*. Munich: Wenschow Franzis Druck GmbH, 1984.

Reeves, Robert. *The Superpower Space Race*. New York: Plenum Press, 1994.

Roth, Ladislav E., and Stephen D. Wall, eds. *The Face of Venus: The Magellan Radar-Mapping Mission*. NASA SP-520. Washington, D.C.: GPO, 1995.

Rubashkin, D. "Who Killed the Grand Tour? A Case Study in the Politics of Funding Expensive Space Science." *Journal of the British Interplanetary Society* 50 (1997): 177–84.

Sagan, Carl. *Billions and Billions: Thoughts on Life and Death at the Brink of the Millennium*. New York: Random House, 1997.

———. *Murmurs of Earth*. New York: Random House, 1978.

———. *Pale Blue Dot: A Vision of the Human Future in Space*. New York: Random House, 1994.

Sagdeev, Roald Z. *The Making of a Soviet Scientist*. New York: John Wiley & Sons, 1994.

Schorn, Ronald A. *Planetary Astronomy: From Ancient Times to the Third Millennium*. College Station: Texas A&M University Press, 1998.

Seamans, Robert C., Jr. *Aiming at Targets*. NASA SP-4106. Washington, D.C.: GPO, 1996.

Space Science Board. *Outer Planets Exploration 1972–1985*. Washington, D.C.: National Academy of Sciences, 1971.

————. *Planetary Exploration: 1968–1975*. Washington, D.C.: National Academy of Sciences, 1968.

————. *Priorities for Space Research 1971–1980*. Washington, D.C.: National Academy of Sciences, 1970.

————. *A Strategy for Exploration of the Outer Planets: 1986–1996*. Washington, D.C.: National Academy Press, 1986.

Swift, David W. *Voyager Tales: Personal Views of the Grand Tour*. Reston, Va.: American Institute of Aeronautics and Astronautics, 1997.

Terzian, Yervant, and Elizabeth Bilson, eds. *Carl Sagan's Universe*. Cambridge: Cambridge University Press, 1997.

"Topics of the Times." Editorial. *New York Times*, 18 January 1920.

TRW Systems Group. *Pioneer: To Jupiter and Beyond*. Compilation of papers from the American Astronautical Society and the American Institute of Aeronautics and Astronautics. Los Angeles: TRW Systems Group, 1971.

von Braun, Wernher. "Man on the Moon: The Journey." *Collier's*, 18 October 1952, 52–59.

von Braun, Wernher, and Cornelius Ryan. "Can We Get to Mars?" *Collier's*, 30 April 1954, 22–29.

Weaver, Kenneth F., with paintings by Ludek Pesek. "Voyage to the Planets." *National Geographic* 138, no. 2 (August 1970): 147–93.

Weissman, Paul R., Lucy-Ann McFadden, and Torrence V. Johnson, eds. *Encyclopedia of the Solar System* San Diego: Academic Press, 1999.

Whipple, Dr. Fred L. "Is There Life on Mars?" *Collier's*, 30 April 1954, 21.

Winter, Frank H. *Rockets into Space*. Cambridge: Harvard University Press, 1990.

Yeates, C. M., T. V. Johnson, L. Colin, F. P. Fanale, L. Frank, and D. M. Hunten. *Galileo: Exploration of Jupiter's System*. NASA SP-479. Washington, D.C.: GPO, 1985.

Unpublished Sources

Hearth, Donald P., Donald G. Rea, William E. Brunk, and Robert S. Kraemer. "Planetary Program Review." Presentation to Thomas O. Paine and NASA General Management, 11 July 1969.

Kochendorfer, Fred D., and Charles F. Hall. "The Pioneer F and G Missions." AAS 71-101. Paper presented at the seventeenth annual meeting of the American Astronautical Society, Seattle, Washington, 28–30 June 1971.

Kraemer, Robert S. NASA notebooks, 1967 to 1990.

———. "Objectives of Planetary Exploration." NASA internal paper, November 1968.

———. "Outer Planets Working Group Minutes, Reports, and Chairman's Notes, 1969–1970."

———. "Solar System Exploration: A Strategy for the 1970s." AAS 71-100. Paper presented at the seventeenth annual meeting of the American Astronautical Society, Seattle, Washington, 28–30 June 1971.

———. Statement before the House Subcommittee on Space Science and Applications, 16 February 1971.

Newell, Homer E. "Relations with the Science Community and the Space Science Board." NASA memo to James C. Fletcher, 3 December 1971.

Nixon, Richard M. "Space Program Statement by the President." 7 March 1970.

"Space Beyond the Outer Planets." Paper presented at the Pioneer 10 Symposium, NASA Headquarters, Washington, D.C., 3 March 1997.

Voyager Program: Mission Operation Report. NASA Office of Space Science Report S-802-77-01/02, 1977.

Interviews

Brunk, William. Interview with author. Annapolis, Maryland. 21 November 1997.

Casani, John. Interview with author. JPL, Pasadena, California. 27 March 1998.

Donahue, Thomas M. Telephone interview with author. 7 February 2000.

Donlan, Charles. Conversation with author. Washington, D.C. 11 June 1997.

Friedman, Louis. Interview with author. Pasadena, California. 28 October 1996.

Giberson, Walker E. "Gene." Interview with author. Pasadena, California. 28 October 1996.

Hall, Charles F. Interviews with author. Washington, D.C. 3 March 1997. Los Altos, California. 22 May 1997.

Herman, Daniel H. Telephone interview with author. 7 November 1997.

Hinners, Noel W. Conversation with author. Potomac, Maryland. 4 October 1997.

Jakobowski, Walter. Interview with author. South Bethany, Delaware. 15 May 1998.

McElroy, Michael B. Telephone interview with author. 30 January 2000.

Naugle, John E. Interview with author. North Falmouth, Massachusetts. 29 July 1997.

Ousley, Gilbert. Telephone interview with author. 29 June 1997.

Parks, Robert J. Interview with author. Balboa Island, California. 22 October 1996.

Pieper, George. Conversation with author. Annapolis, Maryland. 21 November 1997.

Rea, Donald G. Interview with author. 30 April 1997.

Schneiderman, Daniel. Telephone interview with author. 26 and 28 December 1997.

Schurmeier, Harris M. Interview with author. Oceanside, California. 24 October 1996.

Shapley, Willis. Interview with author. Washington, D.C. 11 June 1997.

Soffen, Gerald. Interview with author. Washington, D.C. 24 March 1999.

Trainor, James H. Telephone interview with author. 26 and 27 June 1997.

INDEX